PRO-POOR STRATEGIES IN URBAN WATER PROVISIONING: WHAT KENYAN WATER UTILITIES DO AND WHY THEY DO IT

AKOSUA SARPONG BOAKYE-ANSAH

Promotiecommissie

Promotor:	Prof. Dr. M.Z. Zwarteveen	Universiteit van Amsterdam
Copromotor:	Dr. K.H. Schwartz	IHE Delft
Overige leden:	Prof. dr. I.S.A. Baud	Universiteit van Amsterdam
	Prof. dr. J. Gupta	Universiteit van Amsterdam
	Prof. dr. J.L. Uitermark	Universiteit van Amsterdam
	Dr. M.A. Hordijk	Universiteit van Amsterdam
	Prof. dr. ir. P. van der Zaag	IHE Delft
	Dr. ir. A.R. Mels	VEI Dutch Water Operators
	Dr. S.O. Owuor	University of Nairobi

Faculteit der Maatschappij- en Gedragswetenschappen

This research was conducted under the auspices of the Graduate School for Socio-Economic and Natural Sciences of the Environment (SENSE)

PRO-POOR STRATEGIES IN URBAN WATER PROVISIONING: WHAT KENYAN
WATER UTILITIES DO AND WHY THEY DO IT

ACADEMISCH PROEFSCHRIFT

ter verkrijging van de graad van doctor
aan de Universiteit van Amsterdam
op gezag van de Rector Magnificus
Prof. dr. ir. K.I.J. Maex
ten overstaan van een door het College voor Promoties ingestelde commissie,
in het openbaar te verdedigen in de Agnietenkapel
op donderdag 5 november 2020, te 16:00 uur

door

Akosua Sarpong Boakye-Ansah

geboren te Asokore Mampong

To

Ama; you have made me stronger, better and more fulfilled than I could have ever imagined.

My parents; your love for education brought me this far.

ACKNOWLEDGMENTS

Although only my name appears on the cover of this thesis, a great number of people supported me in different ways towards the accomplishment of this task. While some contributed directly to the research and writing of this thesis, several others provided support and encouragement on this journey. Finding the right words to show my appreciation towards such immense and immeasurable contributions made this section which is not part of the academic work itself the most difficult and exciting section I had to write in the thesis.

Firstly, I would like to express my sincere gratitude to my supervisory team, Professor Margreet Zwarteveen and Dr Klaas Schwartz. I am grateful for your unflinching support and guidance through each stage of this journey, I could not have asked for a better supervisory team. Margreet, I admire your enthusiasm and support throughout this PhD journey. Being a novice to social science research, your vast knowledge in engineering and social science disciplines did not only contribute to the successful completion of this thesis but also to my personal growth as a researcher. Whenever the process becomes overwhelming, you always realised it, and nudge me on by saying "we are almost there"; but we will still work on at least three different versions of whichever paper we are working on; finally, "we got there". Such encouragements helped to broadened my knowledge on the use words in explaining situations in addition to the numbers, graph and charts which I was used to. Klaas working with you these past years on the PhD has been worthwhile. With you, I always had to come prepared and this kept me on track towards a successful completion of the thesis. Your open mentoring approach gave me room to develop my analytical skills and was instrumental in defining the path of my research. One of the biggest constraints towards a successful completion of a PhD thesis is financing, you took control of this aspect and made sure I was financially secured throughout the period, for this, I am extremely grateful. I am looking forward to continue to work with you both, in my new role as a lecturer in the department.

I will forever be thankful to my MSc thesis supervisory team, Professor Pieter van der Zaag, Dr. Maria Rusca and Dr. Giuliana Ferrero. It was just like yesterday when you were guiding me through my research in Lilongwe. Working with you on the Lilongwe case study was the reason why I decided to pursue this PhD. Thank you for inspiring me to succeed in academia.

This research was carried out within the Performance Enhancement of Water Utilities in Kenya (PEWAK) project, led by Vitens Evides International (VEI) and funded by the Netherlands Enterprise Agency. I am extremely thankful for all the support I received from project partners. Leike, thank you for always being there to answers all my questions and introducing me to the right people. To Eddah and the other staff of WASPA thank you for welcoming me into your office. To all the managers and staff of the water utilities I am grateful for granting me unlimited access to information needed to make this thesis successful. I cannot begin to express how grateful I am to all the pro-poor coordinators I interacted with during the fieldwork for this thesis; Kevin, Zaituni, Caro, Jane, Rose, Eric, Winnie, Sally, Irene, Patricia, thank you for your friendship, for being so open with me and all the help you gave me. I am also very grateful to all the people in the various ministries, agencies and organisation (governmental and non-governmental) and residents in the various cities I interviewed, especially, Richard, Peter, Emmanuel, Suzan, Adongo, Abdi for their willingness to speak to me. To Brian Ochieng, Arnold Brian and Daisy thank you for assisting me with interviews and the surveys.

To William Mboya, Damaris and their children; Wendy, Bill and Gabby, thank you for welcoming me into your home and making my stay in Kenya a pleasant one.

I would like to express my gratitude IHE Delft Institute for Water Education, my host institute for the organizational support provided during my PhD studies. I am especially grateful Jolanda, Floor, Jos and Anique for providing me with organizational support. I would also like to offer special thanks to staff and PhD colleagues in the Water Governance department and beyond for their encouragement and friendship over the years. Tatiana I am happy this journey brought us together. Michelle, Herman, Schalke Jan, Jeltjse, Gabriela, Roos, Shahnoor, Adriano, Bossman, Natalia, Irene, Jonathan, Afua, Eva, Nadine, Pedi, Meseret, Jakia, Adey, Jeewa, Mohan, Sida, Zahrah, Mohammed, Mohaned, Raquel, Tesfay, Chalachew, Iman thank you for sharing this journey with me. I am also grateful to Isaac, Imelda and Abdi, whose MSc thesis contributed to this work.

A good support system is important to surviving and staying sane during the PhD process, I was lucky to have so many people around me to provide this support. Naana, George Emelia, Kwabena, thank you for being my family in the Netherlands. Lydia, Jonathan, Koomson thank you for being my go to people whenever I needed to rant about anything. To Wendy and Afua Gyamaah, thank you for taking care of Ama. To Bernice (of blessed memory), Kwaku Minkah, Kwaku Kwarteng, Daniel, Joshua, Justin and all past and present Ghanaian MSc and PhD students of IHE Delft thank for your companionship. Lois, Adelaide, Helena, Jaco, Adeline thank you for your friendship throughout this journey.

During this PhD period, I had the opportunity to worship with the Presbyterian Church of Ghana in Den Haag and the IHE Delft Christian fellowship. I am grateful to past and present members of both fellowships, especially, Taco, Gerda, Miranda, Seth, Marian, Christiana, Esther, Adwoa, Salomey for your warmness and prayers.

I cannot begin to express my gratitude to my family for their love, support, encouragement and prayers throughout this journey. To my parents, thank you for believing in education and giving me the best of it. I am grateful to God for keeping you alive to see this day. I hope I made you proud. To my brothers, the competition between us has pushed me to succeed. To Ama, you coming into my life was the greatest inspiration I needed to bring this work to completion. I had to do this for us.

Last and most importantly, I am extremely grateful to Otwadiapon Kwame who is able to do so much more than we ask or imagine, to Him be the glory forever and ever for his constant Protection, Grace, Mercies and Favour towards me.

SUMMARY

Despite being the region with the fastest growing urban population, Africa has the lowest water supply coverage. The growing urban population mainly resides in peri-urban areas, with houses and residential plots often being illegal or informal and characterized by high population densities. Of the large numbers of people living in these settlements, only about 40 percent have access to clean water, with only 5 percent having access to water from formal water supply networks. The rest of the population resort to small-scale providers and other unimproved water sources for their water needs. Meanwhile, governments of these countries have committed to national and international goals and targets that stipulate universal access. Achieving such goals and targets will mean the provision of sufficient water services to the growing population of the low-income areas. In these cities, public water utilities represent the designated instruments through which governments aim to fulfil this goal of universal access. Meanwhile, in these countries, most water utilities have been turned into commercial entities, who are expected to operate under cost recovery principles.

Given the cost recovery mandate they have to satisfy, water utilities find it challenging to extend services to the low-income areas. A first major challenge stems from the socio-economic and legal characteristics of the consumers living in these areas. Often employed in the so-called informal sector, they not only have low-income levels, but their incomes also fluctuate strongly depending on the availability of work. Because of low and fluctuating income levels, members of poor urban households often find it difficult to pay monthly water bills. The fact that the urban poor often do not have formal land-tenure for their dwellings is an additional reason why water utilities are either reluctant or legally unable to extend networked infrastructures to their houses. A second challenge relates to the growth and development of low-income areas. Most of this growth happens spontaneously and in unregulated ways, which makes it challenging to plan, invest and develop infrastructures for these informal settlements. As a result, water utilities have come to associate low-income areas with widespread illegal water connections, low payment of bills and a high frequency of service disconnection - which results in high volumes of water that cannot be accounted for (non-revenue water) by the utility. Further, the relatively low water consumption rates among consumers in these areas makes service provisioning commercially unattractive, while the recovery of costs is often cumbersome. Indeed, most utility managers do not view low-income areas as a business opportunity, but rather a burden or a risk.

In balancing the obligations of commercial performance and service extension to low-income areas, water utilities have introduced a number of differentiated service delivery strategies. Generally referred to as pro-poor strategies, these entail explicitly distinguishing different water service strategies for different categories of utility customers. The rationale for service differentiation is that water requirements and conditions differ from one neighbourhood to the next. Utilities adopt service differentiation because they realize that they are not equipped to provide the same conventional services to consumers whose water needs and ability to pay for services are very different. Under service differentiation, water services strategies, are tailored to the needs of distinct customer groups through specific technological, financial and organizational systems.

The main objective of this thesis is to understand why and how pro-poor water strategies are employed and implemented by commercialized urban water utilities. While there are a number of studies on the topic, much of this is carried out by international science-policy advisers, many of whom are more interested in discussing or prescribing how pro-poor strategies should be carried out rather than in studying how they are actually being done. While this literature is very useful in understanding the concept, they provide little insight into how water utilities in their everyday operations try meeting diverse and sometimes conflicting performance objectives. This is the gap I intended to fill through an analysis of three dimensions of pro-poor strategies - technological/infrastructural, organisational and financial arrangements, within the context of water supply in urban Kenya. In this thesis, I thus aimed at producing novel insights into how pro-poor water services are actually done. Therefore, I did not set out to deductively test hypotheses generated from existing theories about the functioning of water utilities or the workings of water provisioning modalities. Instead, I choose an inductive approach inspired by Grounded Theory. The inductive approach allowed for the development of theoretical interpretations that enables debates of general characteristics of the study topic, based on data collected.

The technology used for service extension to low-income areas usually concerns so-called appropriate technologies such as water kiosks, yard taps or pre-paid meters. Consumers access water from these technologies through jerry cans or containers. As a result, consumption levels from these technologies are relatively low when compared to consumption from 'normal' networked supply. Moreover, consumers pre-pay for the water they consume, meaning they are not confronted with hefty bills at the end of each month. In resorting to appropriate technologies for water supply in the low-income areas, water utilities frequently engage intermediaries in the service provision process. The utility then provides bulk water through these technologies to the intermediary who distributes the water to individual consumers. Typically, intermediaries are drawn from (1) the informal operators who have prior knowledge of water supply in these areas and (2) the landlords on whose lands the technologies are constructed. Partnering with intermediaries allows the formal utility to maintain a monopoly over water services whilst delegating the riskier water service activities to the intermediaries.

Although it may be beneficial for a water utility as it greatly reduces the risks of providing services in low-income areas, my research findings show that the use of intermediaries is not without concerns. As indicated above, the intermediaries are often drawn from established informal operators who were previously engaged in water supply activities in the low-income areas. In Kenya, the rowdy and exploitative nature of these informal operators had earned them the name Water Cartels. Members of these cartels were the ones responsible for creating artificial shortages, through the vandalism of utility pipes and controlling the local water market. By partnering with them as intermediaries, the utility seeks normalize and control the behaviour of cartel members, thereby reducing the chances that they will "frustrate" their efforts of supplying water to low-income areas. Having being accorded a "legalized" role through their engagement as intermediaries, cartel members obtain a position to exploit low-income consumers by charging excessive prices for poor quality services. Although the use of intermediaries is perhaps necessary in order to provide services, it is equally important to ensure that these intermediaries do not exploit the consumers they are to serve.

Although the water utility is usually the main instrument for achieving universal service coverage, it is not the only actor with an interest in providing water services to low-income areas. In addition to the water utility, the sector consists of regulatory agencies that influence the water services strategies that the utilities develop to service low-income areas. Moreover, (inter)national donors also manifest themselves in the water services sector with the aim of achieving their development objectives. Although these donor organizations have a shared objective of achieving universal service coverage, they often also have secondary objectives that influence how they opt to contribute to this objective. Notably, donors increasingly want to link development aid to international trade. For example, the WASH policy of the Dutch Ministry of Foreign Affairs stipulates in its WASH Strategy 2016-2030 that the Dutch government has committed itself "to provide 50 million people with access to sustainable sanitation and 30 million people access to sustainable drinking water by 2030" (Ministry of Foreign Affairs of the Netherlands, 2016) For water utilities, this means that while accessing donor funding they need to ensure that the donor interests align with the interests of consumers in low-income areas. In various Kenyan cities, technologies delivered through donor-funding have failed to provide the services for low-income consumers.

It is often assumed that 'appropriate technologies' will be more suitable for consumers in low-income areas. However, water services strategies not only concern technological choices, but also an organizational system through which the technology is operated and maintained and a financial system that charges consumers for the water they access. The degree to which a technology is appropriate for consumers in low-income areas is strongly determined by the organizational and financial systems through which the technology is employed. Also, in the Kenyan water service sector, examples exist where consumers in low-income areas pay more for water obtained from kiosks or pre-paid dispensers than high-income consumers who obtain water through an in-house connection. It is thus important to monitor how the use of particular water services strategies unfolds for consumers in low-income areas. In the absence of well-directed subsidies and proper monitoring, while the water utilities and other stakeholders get to satisfy their interest, low-income households may be left worse off.

SAMENVATTING

Terwijl Afrika als regio het snelst verstedelijkt ter wereld, is het percentage van wateraansluitingen de laagste ter wereld. De groeiende stedelijke bevolking woont voornamelijk in peri-urbane gebieden, in huizen en op land die vaak op informele of illegale wijze gebouwd zijn. Deze gebieden worden gekenmerkt door een hoge bevolkingsdichtheid. Van de vele mensen die in deze gebieden wonen heeft slechts 40% toegang tot schoon drinkwater. Slechts 5% heeft toegang tot het drinkwaternetwerk. De resterende bevolking is afhankelijk van kleinschalige waterleveranciers en drinkwaterbronnen van slechte kwaliteit. Tegelijkertijd hebben de overheden van de Afrikaanse landen zich verbonden aan internationale en nationale doelstellingen die watervoorziening voor alle mensen voorschrijven. Om deze doelstellingen te bereiken zal voldoende water geleverd moeten worden aan de bevolking in de armere peri-urbane gebieden. In deze steden zijn drinkwaterbedrijven de aangewezen partijen om universele toegang tot drinkwater te bewerkstelligen. Tegelijkertijd zijn de meeste drinkwater bedrijven commerciële organisaties die kostendekkend moeten opereren.

De noodzaak om kostendekkend te opereren maakt het voor veel drinkwaterbedrijven moeilijk om drinkwatervoorziening uit te breiden naar armere delen van steden. Een eerste obstakel betreffen socio-economische en juridische aspecten. Vaak zijn inwoners van deze gebieden werkzaam in de 'informele' sector, waarbij ze niet alleen een laag inkomen hebben, maar ook een inkomen dat sterk fluctueert in relatie tot de beschikbaarheid van werk. Het relatief lage en fluctuerende inkomen maakt het voor veel huishoudens moeilijk om maandelijks hun drinkwaterrekening te betalen. Voorts beschikken huishoudens in deze armere gebieden vaak niet over het eigendomsrecht van de woonpercelen waar ze wonen. Dit vormt een juridisch obstakel voor het drinkwaterbedrijf om de huishoudens op het waternetwerk aan te sluiten. Een twee reden waarom drinkwater bedrijven terughoudend zijn wat betreft de uitbreiding van het waternetwerk betreft de groei en ontwikkeling van deze gebieden. Deze groei en ontwikkeling is vaak spontaan en ongereguleerd. Hierdoor is het lastig voor het drinkwaterbedrijf om het formele water netwerk uit te breiden in deze gebieden. Het gevolg is dat drinkwaterbedrijven dienstverlening aan deze armere gebieden associëren met illegale aansluitingen, achterstallige betalingen, en een relatief groot aantal waterafsluitingen. De illegale aansluitingen en niet betaalde rekeningen resulteren in grote waterverliezen voor de drinkwater bedrijven. Daarnaast consumeren huishoudens in armere wijken relatief minder water dan huishoudens in rijkere wijken. Dit lage consumptieve niveau maakt het leveren van water financieel niet aantrekkelijk, terwijl het innen van de rekeningen vaak erg bedrijfsintensief is. Het is dan ook niet verwonderlijk dat veel managers van drinkwater bedrijven watervoorziening aan armere gebieden niet zien als een zakelijke kans, maar als een bedrijfsrisico.

In het zoeken naar een balans tussen kostendekkend opereren en water leveren aan armere gebieden, hebben drinkwater bedrijven strategieën ontwikkeld waarbij ze verschillende niveaus van dienstverlening onderscheiden. Deze strategieën worden gepresenteerd als zijnde in het belang van arme huishoudens, maar bieden verschillende klantengroepen een verschillend niveau van dienstverlening. De argumentatie voor deze differentiatie berust op het idee dat verschillende wijken in een stad verschillende water behoeftes hebben. Water bedrijven hanteren verschillende leveringsstrategieën omdat ze niet in staat zijn om hetzelfde niveau van dienstverlening aan alle huishoudens te geven door verschil in de hoeveelheid water dat wordt afgenomen en de mogelijkheid hiervoor te betalen. Door te differentiëren met betrekking tot de leveringsstrategieën wordt de technologie, organisatie en het vaststellen van tarieven en dienstverlening op maat gemaakt voor de klantengroep in een bepaalde wijk.

Het doel van dit onderzoek is om een beter begrip te krijgen van het gebruik van deze strategieën door commerciële drinkwaterbedrijven om armere wijken te bedienen. Hoewel er een aantal onderzoeken naar deze strategieën zijn gepubliceerd betreft het veelal voorstellen van internationale beleidsadviseurs die willen voorschrijven wat er zou moeten gebeuren. Aan de vraag hoe deze strategieën daadwerkelijk in de praktijk worden ingezet is veel minder aandacht geschonken. Hoewel de bestaande literatuur nuttig is om het differentiëren van leveringsstrategieën beter te begrijpen op conceptueel niveau geeft het weinig inzicht in hoe drinkwaterbedrijven in hun dagelijkse werkzaamheden proberen om hun, soms conflicterende, doelstellingen te halen middels deze strategieën. In dit proefschrift beoog ik deze leemte te vullen, door deze strategieën voor drinkwatervoorziening in Keniaanse steden op drie dimensies te analyseren: technologisch/infrastructureel, organisatorisch, en financieel. Hierbij heb ik als doel nieuwe inzichten te presenteren die laten zien hoe leveringsstrategieën voor arme wijken in de praktijk uitgevoerd worden. Het proefschrift beoogt niet om geformuleerde hypotheses van bestaande theorieën te toetsen. In plaats daarvan hanteer ik een inductieve benadering die geïnspireerd is door 'Grounded Theory'. Deze inductieve benadering maakt het mogelijk om, op basis van verzamelde data, theoretische interpretaties te ontwikkelen die discussies mogelijk maken met betrekking tot het bestudeerde fenomeen.

De technologie die gebruik wordt voor levering aan armere gebieden betreft meestal zogenaamde 'geschikte' technologieën zoals water kiosken, tuinaansluitingen, en collectieve aansluitingen waarbij vooruit betaald dient te worden. Consumenten kunnen met jerrycans water ophalen bij deze aansluitingen. Een gevolg hiervan is dat de hoeveel water die geconsumeerd wordt relatief laag is in vergelijking met consumptie van een huishouden met een huisaansluiting. Bovendien moeten consumenten vooruit betalen wanneer ze water consumeren. Hierdoor worden ze niet aan het eind van de maand met een grote rekening geconfronteerd. Bij het leveren van water, gebruiken drinkwater bedrijven vaak tussenpersonen. Het drinkwater bedrijf levert dan bulkwater aan een individu of groep, die vervolgens het water doorverkopen aan consumenten. Vaak worden deze tussenpersonen geworven onder 1) individuen en organisaties die al eerder (informeel) water leverden in deze gebieden en 2) de huiseigenaren op wiens woonperceel de aansluiting wordt aangelegd. Via deze constructie is het drinkwater bedrijf in staat om diens monopolypositie te handhaven terwijl veel operationele risico's worden gedelegeerd aan de tussenpersonen.

Hoewel het gunstig is voor een drinkwater bedrijf om de risico's van dienstverlening in arme gebieden te beperken, laten mijn onderzoeksresultaten zien dat de rol van de tussenpersonen ook een rede tot zorg is. Zoals eerder aangegeven worden de tussenpersonen gerekruteerd uit individuen en groepen die al eerder informeel water leverden aan armere gebieden. In Kenia staan deze individuen en groepen erom bekend hun klanten uit te buiten en worden hierom ook aangeduid als 'waterkartels'. Leden van de kartels hebben in het verleden kunstmatige tekorten gecreëerd door levering van het drinkwater bedrijf te saboteren. Hierdoor konden ze de lokale water markt controleren. Door deze kartels nu als tussenpersonen aan te stellen hopen de drinkwaterbedrijven het gedrag van de waterkartels kunnen normaliseren en controleren. De gedachte hierachter is dat in deze situatie de kartels minder snel geneigd zijn de drinkwatervoorziening te saboteren. Echter, door hun formele rol in drinkwatervoorziening, hebben de tussenpersonen ook een mogelijkheid om consumenten in arme gebieden af te persen door hoge prijzen voor water te vragen. Vaak is het water dat wordt geleverd ook nog van slechte kwaliteit. Hoewel de inzet van tussenpersonen wellicht nodig is om water leveren, is het evenzeer belangrijk om te bewerkstelligen dat deze tussenpersonen de consumenten die zij moeten bedienen niet misbruiken.

Hoewel drinkwaterbedrijven het belangrijkste instrument zijn om er voor te zorgen dat iedereen toegang heeft tot water, is het niet de enige actor die er belang bij heeft dat er water wordt geleverd in armere gebieden. Naast het drinkwaterbedrijf, kent de watersector organisaties die de sector reguleren en de strategieën die gehanteerd worden door de drinkwater bedrijven beïnvloeden. Voorts zijn er nationale en internationale donoren actief die zich gecommitteerd hebben aan gestelde ontwikkelingsdoelen. Hoewel, deze donoren een gezamenlijk belang hebben om dienstverlening uit te breiden, hebben ze ook secundaire doelen die beïnvloeden hoe ze aan dat belang zullen bijdragen. In toenemende mate proberen donoren ontwikkelingssamenwerking aan internationale handel te koppelen. Het Nederlandse ontwikkelingssamenwerking beleid met betrekking tot drinkwater en sanitatie stelt dat de overheid zich committeert om tussen nu en 2030 50 miljoen mensen toegang te verlenen tot sanitatie en 30 miljoen mensen tot drinkwater. Voor drinkwater bedrijven betekent dit dat ze de belangen van donoren moeten behartigen als ze toegang willen krijgen tot de donorfondsen. In verschillende Keniaanse steden hebben door donors -gefinancierde projecten niet geleid tot betere dienstverlening voor armere huishoudens doordat de technologie slecht functioneerde.

Er wordt vaak beargumenteerd dat 'juiste technologieën' beter passen bij de wensen en mogelijkheden van consumenten in arme gebieden. Echter, drinkwatervoorziening is niet alleen afhankelijk van technologie, maar ook van de wijze waarop die technologie toegepast wordt, onderhouden wordt, en de manier waarop de rekeningen geïnd worden. Of deze technologie past bij de behoeften en mogelijkheden van de consumenten wordt sterk beïnvloed door de financiële- and organisatiesystemen die worden gebruikt om de technologie te operationaliseren. In verschillende arme wijken van Keniaanse steden zijn voorbeelden te vinden van consumenten die meer betalen voor hun water dat ze krijgen van water kiosken, dan consumenten in rijkere wijken betalen voor het water dat ze krijgen via hun eigen huisaansluiting. Het is dus van belang om opmerkzaam te blijven over hoe systemen voor dienstverlening op het gebied van drinkwatervoorziening zich ontwikkelen in arme wijken. Want alhoewel, het drinkwaterbedrijf en donoren wellicht hun doelstellingen halen, is de consument met een laag inkomen, door het ontbreken van gerichte subsidies en adequate regulering, wellicht degene die de hoogste prijs betaalt.

CONTENTS

1

INTRODUCTION

"I have lived here since the 1970s and access to clean piped water has always been a challenge. The government has tried a number of interventions to get us connected to the network, these interventions work for a while and then they fail. This current Delegated Management Model seems to be working, we hope it can be maintained"
(resident, Nyalenda)

This is the reply I received from a septuagenarian living in Nyalenda, the largest informal settlement in Kisumu, Kenya. I had asked him about the water supply situation in the area. Over the years, residents in this informal settlement have had to contend with limited access to water of improved quality, and have been confronted with a number of more or less successful interventions. This situation is not unique to residents in Nyalenda: many informal, low-income settlements which serve as the home to the poor in cities in developing countries experience similar challenges in accessing water. In these cities, only about 40 percent of the population living in low-income areas (LIAs) have access to clean water, with only 5 percent having access to piped water on their premises (Dagdeviren and Robertson, 2009). The remainder of the population rely on small-scale water providers and other unimproved water sources to satisfy their water needs (Kariuki and Schwartz, 2005; Ahlers et al., 2014).

These challenges have not gone unnoticed by governments and development practitioners. Efforts towards improving access to water in such marginalised areas have featured prominently on development agendas – from the era of the "International Drinking Water Supply and Sanitation Decade" (1980-1990) the goal of which was to improve access to about 1.8 billion un/under-served people to this current era of the Sustainable Development Goals (SDGs) which is targeting some 2 billion people who still lack access to safely managed drinking water (O'Rourke, 1992; WHO and UNICEF, 2017). The interventions specifically targeted at improving access to poor groups have evolved over the years. Referred to as pro-poor interventions, these aim to provide the poor with access to clean water in enough quantities and at affordable rates (Cross and Morel, 2005; Berg and Mugisha, 2010; Ryan and Adank, 2010)

In this thesis, I critically examine the concept of pro-poor water services. Taking water supply in urban Kenya as the empirical point of departure, the thesis focuses on how urban water utilities mobilize and implement pro-poor strategies. The thesis was carried out within the Performance Enhancement of Water Utilities in Kenya (PEWAK) project, led by Vitens Evides International. Through benchmarking, collective learning and innovative financing, the project, among other goals, sought to close the gap in access to water services between poor and richer households and create an enabling environment for water service providers and utilities to develop a strong focus on service delivery to the urban low-income areas.

1.1 EVOLUTION OF PRO-POOR SERVICES

1.1.1 The International Drinking Water Decade and the Modern Infrastructural Ideal (MII)

At the 1977 Water Conference in Mar del Plata, the United Nations declared the 1980s (1981-1990) the "International Drinking Water Supply and Sanitation Decade". By this declaration, the international community sought to provide clean water and sanitation for all people throughout the world by the end of the decade. The Decade focused on "rapid coverage", so for most countries, this meant the continuous development of water infrastructure to un/under-served areas (O'Rouke, 1992; Luageri, 1986).

During this decade, the development of water services followed the Modern Infrastructural Ideal (MII). The MII entails "monopolistic, integrated and standardized service provision" (Schwartz and Sanga, 2010: 765). It emphasises the availability of one standardized technology, by a single utility provider (often public) for all citizens in a given service area (Graham and Marvin, 2001; Jaglin, 2008). Following this ideal, pro-poor services implies the extension of networked services to LIAs and households by the same water provider and through the same uniform and standardised network that serves other parts of the city and other households. It also means the standardized use of in-house connections for water supply, also in the LIAs. The ideal implicitly incorporates the use of cross-subsidies and different tariff structures, with middle and high-income consumers subsidizing the cost of water services for the low-income consumers.

The decade however failed to achieve its goal, as the number of people lacking access to safe water was almost similar when the decade ended as compared to when it began (Schwartz, 2006). The standardisation that the MII entailed often led to an improvement of services to the already served areas instead of service extensions to the un/underserved areas (Laugeri, 1986; O'Rourke, 1992).

1.1.2 Institutional Development and the Introduction of Appropriate Technologies

By the end of the 1980s, water supply systems in most cities of the developing world were facing growing problems of quality, reliability, and coverage. Many utilities, according to the prevailing problem diagnosis of the time, were dominated by a civil engineering culture: they were more interested in building large hydraulic works than in operating them. Subsequently, in the early 1990s, the international community started emphasising the importance of managerial and institutional solutions to problems of water provisioning. In parallel with changes in the world's political order, the new paradigm also emphasised private sector participation and the reduction of government involvement in the sector. The implementation of such privatization principles became a conditionality for accessing loans and benefits from the international financial organisations such as the World Bank and the IMF (Amenga-Etego and Grusky, 2005; Bohman, 2012).

During this era, governments and the international community also gradually moved away from the ideal that water services should be the same for all people regardless of their socioeconomic circumstances, to instead accept differences between people and communities based on location and income. Hence, the idea emerged that low-income communities should access water through so-called appropriate technologies. Defined as "that process or technique which provides a socially or environmentally acceptable level of service or quality of product at the least social cost" (Gunnerson, 1978; cited in O'Rourke, 1992), the adoption of these technologies entailed the acknowledgment by water service providers that the only way to extend water to low-income communities was to differentiate services. The use of appropriate technologies was not only to reduce the cost of water, but also to install devices whose operations and maintenance would make use of existing local knowledge (O'Rourke, 1992). However, the idea of using existing local knowledge was not followed through to the implementation stage. Authorities accepted[1] and installed pre-designed devices that had 'worked elsewhere'. In many cases, these devices stopped functioning after they had been in use for only a short while (O'Rourke, 1992). A more detailed discussion on how appropriate such technologies are follows in the ensuing chapters of this thesis.

1.1.3 Commercialisation and the Promise of Universal Access

It soon became apparent that in most developing countries, the massive investment needed in the sector coupled with low tariffs rendered full privatisation unattractive to multinational companies and investors (Dagdeviren, 2008; Kirkpatrick and Parker, 2006; Lobina, 2005). Many countries, including Kenya, therefore shifted from full privatisation

---

[1] These technologies/devices were handed down by donors and financial organisations.

to the commercialisation of water services as the "de facto policy" (Dagdeviren, 2008; Prasad, 2006; Smith, 2006). Commercialization does not necessarily entail the involvement of private or non-state actors (Furlong, 2015; Van Rooyen and Hall; 2007, Bakker, 2007; Schwartz, 2008; Smith, 2004; 2006), but means that existing public utilities start operating on a commercial or market basis (De Albuquerque and Winkler, 2010), with the state as the only shareholder. Financial and aid agencies (including the World Bank, the IMF and other donor organisations) actively embraced and promoted this market oriented and output-based approach as a good substitute to the long-established bureaucratic model of public management (Schwartz, 2008). The main ingredients of the commercialisation approach are the introduction of corporate and commercial principles and performance contracts (Banerjee and Morella, 2011; Furlong and Bakker, 2010; Furlong, 2010). The approach entails the financial and managerial ring-fencing of state-owned water utilities into "stand-alone business units" (McDonald and Ruiters, 2005), intended to increase the level of autonomy and limit political interference (Smith and Hanson, 2003). Proponents had high hopes that commercial models would help make the resources available that utilities needed to make infrastructural investments, while improving the efficiency of operations as well as extending services to all, including the [poor] population living in the under or unserved [low-income] areas and at "affordable prices" – prices that will not deter the poor from accessing the service (Yarrow, 1999; World Bank, 1992; Shirley, 1999; Klein, 1996; Freire and Stren, 2001). However, the two billion people still lacking access to improved water supply at the inception of the SDGs show that commercialisation had not produced all the hoped-for benefits, especially in terms of making water accessible for the urban poor. This realisation prompted governments, developmental practitioners as well as donors to consider new ways of getting the urban poor connected to improved water sources (Gerlach and Franceys, 2010:1230).

The question of how commercialised water utilities mobilize and implement pro-poor water services is central to this thesis. There is currently a growing literature on this topic. Yet, much of this is written from the perspective of (often international) policy practitioners who are more interested in discussing or prescribing how pro-poor strategies of water services provisioning should be done than in studying how they are actually being done (Berg and Mugisha, 2010; Cross and Morel, 2005; Ehrhardt et al., 2007; Mara and Alabaster, 2008; Marson and Savin, 2015; Tre´molet and Hunt, 2006). Studies discussing how pro-poor strategies are actually implemented, how they work in practice or accounts of the everyday practices of the implementing agents (the water utilities) are few. This is the gap that I intend filling with this thesis.

1.2 RESEARCH OBJECTIVES

The main aim of this thesis is to investigate pro-poor water strategies and their use by water utilities, examining how the provisioning of services to urban low-income areas happens. With this topic, the thesis touches on how commercial urban water utilities always need to balance goals of cost recovery with those of universal access. The possible tensions between the two goals form a much-debated topic in the literature. I do not intend to dwell so much on these debates. Instead, the thesis focuses on how and why water utilities identify, select and implement pro-poor strategies. I analyse three dimensions of pro-poor strategies – technological/infrastructural, organisational and financial arrangements - within the context of water supply in urban Kenya to explore the everyday experiences and practices of urban water utilities in their quest to supply water to the urban poor. In doing so, I also aim to shed light on the institutional environment which the utilities navigate in order to carry out their mandate. To reach this objective, the following research questions were formulated:

1. Why do utilities select and develop specific pro-poor interventions?

With this question, I aim at elaborating on the argumentation for the promotion of pro-poor water services, how the concept is understood and used from the perspective of the managers of water utilities.

2. How do utilities implement pro-poor water strategies?

Here I elaborate on the three dimensions of pro-poor water services, what they entail and how they are used by utilities to satisfy their water supply mandate in the low-income areas.

3. How does the adoption of pro-poor strategies impact water supply in urban low-income area?

This question examines the effects of the interventions on the operations of the implementing water utilities. By examining different dimensions of access, the question also probes into the appropriateness of the intervention for consumers in low-income areas.

1.3 METHODOLOGY

1.3.1 Research Strategy

To explore and shed light on how and why pro-poor interventions are designed and implemented by water utilities, I use a case study approach as this helps to start understanding how certain decisions are made, how they are implemented and the outcomes that are generated (Yin, 2009; Hancock and Algozzine, 2006). Because the intent of this study was to produce novel insights into how pro-poor water services are actually done, I did not set out to deductively test hypotheses generated from existing theories about the functioning of water utilities or the workings of water provisioning modalities. Instead, I chose an inductive methodological approach inspired by Grounded Theory (GT). I intentionally wanted to allow for new understandings to emerge, which is why I deliberately refrained from adopting the theories about the behaviour of utilities assumed in the international science-policy domain, a domain dominated by donors. The inductive approach allowed for the development of new theoretical interpretations that enable debates of general characteristics of the study topic, based on data collected. My approach thus consists of the conceptualization of insights based on what emerges from data collection (Figure 1). A definition of GT is "the discovery of theory from data systematically obtained and analysed in social research" (Glaser and Strauss, 1967). It is "an inductive theory discovery methodology that allows researchers to develop a theoretical account of the general features of a topic while simultaneously grounding the account in empirical observations or data" (Glaser and Strauss, 1967; in cited Martin and Turner, 1986: 141).

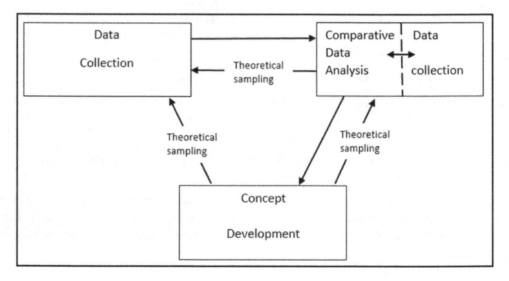

Figure 1: Grounded theory methodological framework.
Source: Author; adapted from Locke, 1996.

1.3.2 Doing the Research

As earlier mentioned, this thesis forms part of the PEWAK project, which involves nine urban water utilities and one rural water utility in nine different counties in Kenya. In line with the GT methodology, the research was carried out in four phases.

The first phase involved an initial visit to the study area in the early parts of 2016, which allowed me to familiarise myself with the objectives of the entire project, identify the areas of interest for my thesis and select the cases (utilities) for the research. It also included the review of relevant literature which helped in the identification of the problem and subsequently, the research objectives and questions. The reviewing of literature continued through the different phases of the research. I selected three water utilities, Kisumu Water and Sanitation Company (KIWASCO) in Kisumu County, Kericho Water and Sanitation Company (KEWASCO) in Kericho County and Nakuru Water and Sanitation Company (NAWASSCO) in Nakuru County for in-depth study. This selection was based on:

- The existence of different forms of pro-poor water interventions;

- The different administrative boundaries within which the utilities exist and operate;

- Ease in accessing secondary data on the basis of previous studies carried out in these study areas;

- Language; the fact that English is widely spoken in the three cities.

Further, all three utilities had dedicated units within the organisation managing the pro-poor interventions in the low-income areas. This was especially important because, managers of these units were responsible for monitoring the activities of the water utilities in the LIAs. However, unlike KIWASCO and NAWASSCO which had well established interventions as well as management units (created in 2009 and 2012 respectively), the interventions as well as the management units in KEWASCO were fairly new; established in 2015.

The second phase of the study involved the collection of data from the three selected cases. Fieldwork for this phase was carried out in 2017, with interviewing and participant observation being the main data collection methods. During this phase, interviews were conducted in an informal and semi-structured manner. The semi-structured approach provided the space for a thorough exploration of questions in a less flexible manner. Respondents were drawn through snowball sampling, from different stakeholders involved in pro-poor access and interventions. The main initial sources of information /data at this stage of the research were the managers or coordinators of the pro-poor departments of the selected utilities. The offices of these coordinators were the first port of call on my visit to the utilities. They were also the people I went to should I need confirmation of data or introduction to other respondents. In addition to the pro-poor coordinators, information was obtained from other staff from the water utilities and from related governmental departments and ministries, NGOs, CBOs, consumers in the LIAs and donor organizations. In my interactions with the coordinators, and other respondents, my own experience of working with a water utility (Ghana Water Company Limited) came in very handy. This experience allowed me to identify with the respondents (especially the utility staff) and gain their trust. They were forthcoming with information and to some extent, they became co-researchers, pointing out and adding relevant information I missed in my interactions with them. As a result of this trust, I got invitations to attend meetings where pro-poor interventions were on the agenda. The semi-structured interviews were complemented with field observations of the daily practices, activities and culture within the water utilities, NGOs, CBOs, among consumers, government officials and other stakeholders.

As the research progressed into the third phase (which also involved further data collection), I employed a more structured and formal interviewing approach and also conducted surveys with structured questionnaires. I also intended to do focus group discussions among members of CBOs and consumer representatives, but had to abandon this research method because respondents were not comfortable expressing their views in the presence of others.

The fourth and final phase of the study involved the comparative analysis of data collected in the previous three phases and further data collection to "fill in the gaps". Results from the various phases of the research were disseminated into scientific articles, which form the different chapters of this thesis.

2
LITERATURE REVIEW[2]

In dealing with the challenge of providing water services to urban low-income areas, the concept of 'pro-poor water services' is popular in the policy literature. Based on an extensive literature review, this article examines the relation between the implementation of pro-poor water services and the equity of access. Pro-poor water services comprise a set of technological, financial and organisational measures employed by utilities in developing countries to improve service provision to low-income areas. In practice, the combination of low-cost technologies, which limit consumption, measures to enforce payment for services, and the use of community-based and private suppliers, means that pro-poor service often entails the utility delegating part of the responsibilities, costs and risks of providing services to those living in low-income areas. Indeed, it is by partially withdrawing from these areas that utilities succeed in reconciling the objective of improving service delivery with the realisation of their commercial objectives. Our analysis shows that in implementing pro-poor service delivery strategies, there is a risk that concerns about cost recovery and risk reduction on the part of the utility prevail over those about the quantity, quality and affordability of the service for the poor.

[2] This chapter is published as: Boakye-Ansah, A. S., Schwartz, K. and Zwarteveen, M. (2019). Unravelling pro-poor water services: what does it mean and why is it so popular? Journal of Water, Sanitation and Hygiene for Development, 9, 187-197.

2.1 INTRODUCTION

Past efforts of development agencies and governments to improve access to potable water have paid off: 96% of the world's urban populations now have access to improved water sources (WHO and UNICEF, 2015). Not all urban dwellers, however, benefit from this improvement. Most notably, in low-income areas of cities in Sub-Saharan Africa, only 64% of the population have access to improved water sources and a mere 5% access water through sources located on their residential premises. Viewed from the perspective of the service provider, the task of providing water services to low-income areas is challenging. A first major challenge stems from the socio-economic and legal characteristics of the people living in these areas. Often employed in the so-called informal sector, they not only have low income levels, but their incomes also fluctuate strongly depending on the availability of work. Because of low and fluctuating income levels, members of poor urban households often find it difficult to pay monthly water bills (Berg and Mugisha, 2010: 592). The fact that the urban poor often do not have formal land tenure for their dwellings is an additional reason why water utilities are either reluctant or legally unable to extend networked infrastructures to their houses.

A second challenge utilities face when attempting to provide water service to the poor relates to how low-income areas grow and evolve. With an urban growth rate of 4.58% and a growth rate of 4.53% for low-income areas, in Sub-Saharan Africa "slum growth is virtually synonymous with urbanization" (UN-Habitat, 2006: 22). African informal settlements house an estimated 50% - 60% of the urban population (Cross and Morel, 2005: 52). Yet, most of this growth happens spontaneously and in unregulated ways, which makes it challenging to plan, invest and develop infrastructures for these informal settlements. This combination of reasons explains why water utilities have come to associate low-income areas with high rates of illegal connections and low rates of bill payment (Heymans et al., 2014: 3). A proliferation of illegal connections in turn produces high-levels of unaccounted-for water (Castro and Morel, 2008: 291). Further, the relatively low water consumption rates among consumers in these areas makes service provisioning commercially unattractive, while the recovery of costs is often cumbersome. Indeed, most utility managers do not view "low-income areas as a business opportunity, but rather a burden or a risk" (Heymans et al., 2014: 3). Water utilities prefer to focus on areas which are less risky and that allow for easier and more secure cost-recovery: middle and high-income areas where consumption rates are (expected to be) higher (Castro and Morel, 2008).

In attempts to reduce risks while sustaining commercial viability, utilities are resorting to a range of distinct service delivery strategies for low-income areas (Water and Sanitation for the Urban Poor (WSUP), 2015: 5), often referred to with the term 'pro-poor' water services. Itself, the concept of pro-poor has been around for a few decades (Komives, 1999). Yet is relatively recent that donors and government organizations have started to use the term for differentiating water services provision between low- and higher income areas. Articles and policy documents speak of 'pro-poor strategies' (Cross and Morel, 2005; Berg and Mugisha, 2010), 'pro-poor units' (Water and Sanitation Programme (WSP), 2009a), 'pro-poor water service delivery' (Ryan and Adank, 2010), 'pro-poor technologies' (Paterson et al., 2007), 'pro-poor water provision' (Appleblad Ferdby and Nillson, 2013), 'pro-poor WASH Governance' (UN-Habitat, 2016), 'pro-poor water governance'(Connor, 2005), 'pro-poor utilities' (WSP, 2009a) and even 'pro-poor private participation' (MacGrannahan and Sattherthwaite, 2006). In general, these documents employ the term pro-poor to refer to a form of service provision specifically designed to alleviate the barriers that water consumers in low-income areas face when accessing water services (Mason et. al., 2016): prohibitive costs, the physical location of infrastructure and the regulation of services. Many utilities in the co-called Global South have established dedicated 'pro-poor units' or departments, which have the specific responsibility to provide services to low-income areas (WSP, 2009a). Funding for these units or departments may come from different sources, and may be organized differently from that of 'normal' service provision. Underlying the establishment of pro-poor units is the conviction that serving low-income areas requires distinct strategies and a targeted commitment to implementing these strategies (WSUP, 2015: 7).

In this paper, we shed light on the concept of pro-poor services in urban water provisioning, focusing in particular on how the implementation of service differentiation entails changes in the equity of access. Our interest in this emerges from a wider discussion in the water management and governance literature on the importance and meaning of equity in domestic water provision (see for instance Goff and Crow, 2014) – a discussion that is importantly prompted by concern about how the so-called neo-liberal turn in water governance prioritizes profits over people (Bakker, 2002), but that was also energized by the explicit adoption by international agencies like the WHO/UNICEF Joint Monitoring Program (JMP) (WHO, 2011) of equity as a major goal.

We use an extensive review of the relevant literature to, first of all, illuminate how the concept of "pro-poor services" is understood and used within water utilities. The review shows that use of the concept involves and is part of a broader shift in thinking about how responsibilities, costs and risks of water services provision should be organised: from one of (MII) to an acceptance of the co-existence of many differentiated forms of service delivery. We then go on to show how this differentiation also entails a change in thinking about and dealing with equity, as it entails creating and accepting differences in access to services between areas and people. One relatively straightforward way of defining equitable access that serves the purpose of this article is: "access being similar for all people irrespective of where they live, whether they belong to vulnerable or marginalized groups, and to the associated costs being affordable for all users" (UNECE and WHO, 2013).

Our analysis reveals that there is a mismatch between the argumentation to promote pro-poor services and the reasons that explain its popularity: whereas pro-poor services are promoted to better tailor services to the needs of the urban poor, the enthusiasm of utilities to embrace the concept mainly stems from how it allows them to demarcate 'the poor' (or low-income areas) as a distinct customer category. In developing this argument, the article first illustrated how the idea of pro-poor services marks a de facto shift away from the MII. The article then identifies three dimensions of pro-poor services provisioning (the organization, the technology and the financing) and explains how these dimensions can be viewed and are presented as 'pro-poor'. The article continues with a critical examination of the practical implementation of 'pro-poor services', discussing this against idea(l)s of equity.

2.2 FROM MII TO PRO-POOR SERVICES

For a large part of the 20th century, the MII (see Graham and Marvin, 2001) served as the reference against which actual developments in the water services sectors in both developed and developing countries were assessed. The MII has it that services are to be provided through 'public or private monopolies, for singular and standardised technological grids across territories [and] the "binding" of cities into supposedly coherent cities' (Graham and Marvin, 2001: 91). The MII thus implies having a single water operator extend services across the city territory through a standardised and centralised network, with in-house connections for all consumers – irrespective of income or other differences. When following this ideal, serving low-income areas basically means extending the network into these areas. The provider can finance the provision of services to poorer clients either through cross-subsidies and differentiated tariffs – with middle- and high-income consumers subsidising the costs for low-income consumers – or by complementing funds from a nation's budget (Jaglin, 2008: 1905). For a long time, the MII continued to implicitly serve as the norm against which actual water service provision was measured and assessed, even when in many cities in developing countries, it provided a very inaccurate metaphor to describe and make sense of actual water provisioning realities. In recognition of this, Bakker (2003) proposed replacing the metaphor and idea of MII to describe urban water services in developing countries with the metaphor of "archipelagos: spatially separated but linked islands of networked supply in the urban fabric" (see also Schwartz et al., 2015; Furlong and Kooy, 2017).

Also, water sector professionals are increasingly realising that attaining the MII is unrealistic. This is why they have started to call for a new paradigm of water services provision (Mara and Alabaster, 2008: 120), one in which the MII idea and ideal that access and affordability to water should be similar to all customers or clients is being replaced with that of service differentiation (see Jaglin, 2008: 1905). Service differentiation entails explicitly distinguishing between different water service strategies and between different customers or clients. This happens by employing different technologies, adopting variations in the ways in which service provision is organised and by differentiating funding and cost-recovery mechanisms. The rationale for service differentiation is that water requirements and conditions differ from one neighbourhood to the next. Utilities adopt it because of the recognition that they are not equipped to provide the same conventional services to consumers whose water needs are different, while their willingness and ability to pay for services may also differ (Njiru et al., 2001: 277; see also Mara and Alabaster, 2008). It 'reflect(s) a pragmatic move towards accommodating social and spatial disparities in a polarized city' (Jaglin, 2008: 1898)

2.3 PRO-POOR SERVICES: APPROPRIATE TECHNOLOGY, FINANCE AND ORGANISATION

As noted, the idea of pro-poor services rests on differentiating services across spaces and people. More specifically, it signals a set of mechanisms and methods to differentiate services to those consumers who are characterised as poor. In practice, three ways of doing this can be distinguished. First, the technologies used for providing services are adapted to better suit the needs and abilities of the urban poor (appropriate technology). Secondly, the management and organisation of service delivery are adapted from one single provider that covers the whole service delivery and process to multiple, small-scale providers who receive bulk water from the utility. Thirdly, the combination of different technologies and a distinct organisation of service delivery are funded through a separate financial regime, one that caters to the affordability of the service for low-income households. These three dimensions are strongly interlinked. Below we discuss each of them in detail.

2.3.1 Appropriate Technology

The central tenet of appropriate technology is that there needs to be a fit between the technology used for providing services and the users of those services. The concept was first coined and promoted by Schumacher (1973) when he used the term 'intermediate technology' to describe a technology that is 'vastly superior to the primitive technology of bygone ages, but at the same time much simpler, cheaper and freer than the super-technologies of the rich'. In water supply terms, conventional centralised networks with in-house connections can be seen to represent the 'super-technologies of the rich': these are deemed too expensive or difficult to operate for poor consumers. What, then, constitutes a technology that is thought to be (more) appropriate or suitable for users in low-income areas? According to the literature, the first and probably most important characteristic of pro-poor technologies is that they should be affordable: those with low(er) incomes should be able to cover the costs of investment, operation and use (Mara, 2003: 453). Interestingly, in water provision, the affordability argument seems to apply more to the utility than to the consumer. The construction or extension of conventional centralised networks in or to low-income areas is often an expensive endeavour for utilities: the capital costs involved are high, especially when these areas are geographically remote. Especially when they themselves are responsible for recovering the costs of investment, utilities are hesitant to make such an investment, also because chances of recovering these costs here are much lower than elsewhere. The use of low-cost technologies thus allows utilities to provide services to low-income areas with lower commercial risks (Kayaga and Franceys, 2007: 277).

A second important characteristic to make a technology suitable or appropriate for low-income areas has to do with matching it to the distinct socio-spatial characteristics of these areas: high population densities and unplanned growth. Here again, the challenges faced by utilities appear more prominent than those faced by poor consumers (Criqui, 2015: 95; see also Kayaga and Franceys, 2007). The Water Utility Partnership (WUP), (2003), for instance, refers to the haphazard layout and the difficult geographic and environmental conditions that characterise low-income areas. Such conditions require improving the flexibility of the infrastructure for water provision (Gulyani et al., 2005).

A third characteristic of appropriate technologies that is often mentioned focuses on the requirements of (responsibilities for) operation and the maintenance of technology. Where conventional networked infrastructures tend to rely on highly trained utility staff to operate and maintain the network, so-called appropriate technologies often entail a delegation of operation and maintenance responsibilities to local users. This means that technologies need to be easier to operate and maintain: the same skills and expertise but also the same resources and support cannot be assumed to exist among inhabitants of low-income areas (Hunter et al., 2010). The involvement of users in the operation and maintenance of technologies (Solo et al., 1993: 6) has two dimensions. First of all, user involvement requires that the technology is socio-culturally acceptable to users (Mara, 2003) and secondly, the involvement of users requires the technology to fit with the available capacities within low-income areas.

A fourth characteristic links the technology to the amount of water consumed, which is assumed to be lower in low-income areas. This is so mainly because the ease of accessing water is lower as compared to conventional in-house connections, whereas the greater distance between the consumer and the water source (kiosk or standpipe) also often reduces consumption rates (Gleick, 1996). Again, this relatively low per capita consumption may also be 'appropriate' for the utility. First, water utilities may not have enough water to supply low-income communities with similar amounts of water as other ('normal') consumers. Indeed, in many Sub-Saharan countries, water rationing is being implemented as a measure to deal with limited availabilities. In countries like Kenya (Hailu et al., 2011) and cities like Lilongwe (Alda-Vidal et al., 2018) and Accra (Stoler et al., 2013), water utilities deal with this by resorting to rationing schemes. If consumption rates in low-income areas would equal those of users connected to the conventional network in the middle- and high-end areas in cities, the water utility would have even less water to distribute among all consumers. Hence, making use of appropriate technologies allows the water utility to expand services to low-income areas without compromising service delivery to areas already served through conventional networks (Ledant, 2013; Boakye-Ansah et al., 2016; Alda-Vidal et al., 2018; Rusca et al., 2017; Tiwale et al., 2018). Lower consumption rates may also be convenient in view of the limited ability of users in low-income areas to pay for services (Mara, 2003; Berg and Mugisha, 2010). We discuss this in more detail in the next section.

In summary, appropriate technologies are considered 'appropriate' as they require less investment cost, match better with assumed preferences and capacities of poor customers (or residents of low-income areas), are suitable for the topographic and environmental conditions of low-income areas and allow the utility to extend services without compromising services to other areas and customers. While embraced by many utilities, the idea of servicing the poor through so-called appropriate technologies is not without criticisms. One important source of critique is that appropriate technologies create distinctions between more and less deserving customers – with appropriate technologies providing a lower level of service to those in the latter category. The water needs of poor customers or inhabitants of low-income areas are not very different from those of wealthier people, or those residing in better-connected parts of the city. A study of service provisioning in low-income urban areas in Kenya, for instance, found that those consumers who were provided water through water kiosks were also less satisfied with water services (Gulyani et al., 2005). That the consumers are 'forced to rely on them' does not mean that they accept them as satisfactory, or would not be interested in the water provisioning options available to other citizens (Gulyani et al., 2005: 1262). This study questions how 'demand-responsive' and equitable services provided through so-called appropriate technologies really are. A second criticism concerns the low levels of consumption associated with appropriate technologies, as well as with the quality of the water provided. Examples from water kiosks in Lilongwe, Malawi, show that clients of kiosks only consume 14 litres per capita per day (Hadzovic, 2014). A study on consumption from standpipes in Ghana even measured daily water use of only 3 litres per capita (Adank and Tuffour, 2013). In both places, these consumption levels are far removed from the 50 litres per capita per day that the World Health Organization recommends. Besides the quantity of water, the quality of water provided at these water points has also been questioned. A study on the quality of water collected at kiosks in Lilongwe, Malawi found that the water that consumers finally use is contaminated when they have to wait in queues with their vessels and from carrying the filled vessels over usually long distances to their homes. The quality deteriorates even further when the water is stored for future use (Boakye-Ansah et al., 2016; Rusca et al., 2017).

2.3.2 Pro-Poor Finance

From the perspective of the utility, doubts about the ability of inhabitants of low-income areas to pay for water services form one of the main challenges for service expansion to low-income areas. As highlighted in the introduction, the urban poor often struggle to pay connection fees and water bills, as their incomes are low and irregular. This is why utilities frequently associate the provisioning of water to low-income communities with high levels of non-payment (Almansi et al., 2010). Payment revolves around two types of costs, the connection fee and the monthly water bill. Connection fees concern the costs of labour, piping materials, the water meter and other connection expenses. The normal costs of connection to the conventional network are usually unaffordable for consumers in low-income areas, who often have irregular incomes and as such struggle to pay the relatively high lump-sum connection fee (Jimenez-Redal et al., 2014: 23). MacIntosh (2003: 79) notices the high connection fee (often over US$100) as one of the main reasons that the poor are not connected to piped water. Franceys (2005: 211) even suggests that utilities may purposively maintain connection fees so high so as to reduce demand, especially when water availability is a problem. In terms of pro-poor water services, a variety of (combined) measures has been proposed to reduce the burden of the connection fee. First of all, some advocate for the use of micro-credit facilities as a way of allowing low-income households to access funds to cover the connection fee. The second measure concerns the amortisation of connection fees over several years, allowing the lump-sum fee to be spread out over time. A third measure concerns (partly) replacing the cash fee with an in-kind contribution from the consumer (Franceys, 2005), by asking the consumer to contribute to the labour needed for making the actual connection (Almansi et al., 2010).

Once a connection is established, however, the periodic costs of the water bill form another challenge for low-income households. A combination of an overall lower ability to pay with fluctuating income levels makes it difficult for low-income households to pay for monthly water bills (Njiru et al., 2001; Mara, 2003; Berg and Mugisha, 2010). This means that such households face a real danger of being cut off from water supply due to non-payment of bills.

Under the umbrella of pro-poor water services, two ways have been proposed to address the burden of volumetric charges for water. The first concerns the use of subsidies to establish a 'social tariff' for low-income households connected to the conventional network. Such social tariffs may, for instance, take the form of progressive or increasing block tariffs (IBTs): providing water at different charges, depending on the amount of water consumed (in relation to defined blocks). Tariffs charged with IBTs increase as consumption moves from a lower consumption block to a higher consumption block. IBTs have become the standard in developing countries (Boland and Whittington, 1998). Perhaps, the most famous example concerns the life-line tariff implemented in South Africa, which provides 6 m^3 of water per month for free (Peters and Oldfield, 2005; Narsiah, 2010; Renouf, 2016).

The second way of ensuring that consumers are able to pay the volumetric charges for water is by effectively reducing consumption rates through the use of appropriate technologies. As noted, the 'appropriate technologies' used for supplying water to low-income areas reduce consumption levels (and thus require lower payments). Many of them also require payment at the moment of accessing water (for example, payment at a water kiosk or through pre-paid meters). In the narrative of pro-poor services, these technologies appear as devices to relieve people of the financial burden associated with connection fees, deposits and bulk monthly payments (Tewari and Shah, 2003; Berg and Mugisha, 2010). Some even present pro-poor water services through appropriate technologies as promoting financial 'self-sufficiency' and being 'demand-responsive', in that they allow households to pay for water when they use it and in the amounts that they demand. The prepayment for relatively small quantities of water allows users to consume water without risking arrears in bills, which will warrant disconnections. In doing so, the financing scheme that accompanies appropriate technology is seen as better fitting the needs and demands of low-income households (Heymans et al., 2014).

While all this may be so, the financial arrangements associated with social connections, block tariffs and appropriate technologies in practice often do not turn out to be very equitable. A study of social connections in West Africa found that they may be useful to target 'the relatively poor who own property', but that their use for 'the poorest' may be limited (Lauria et al., 2005: 25). This is because such social connections often require land tenure and 'the poorest' are likely to be unable to pay the monthly water bills from an in-house connection. Whittington, (1992), has highlighted the possible adverse effects of IBTs in developing countries. Because often multiple low-income households share a single water connection, the cumulative consumption of these households can place the shared connection in the higher tariff blocks. 'In such situations IBT structures may actually have the opposite effect than the intended equity objective; they may penalize low-income households instead of helping them' (Whittington, 1992: 72). Finally, although low consumption levels may lead to lower payments for water, in comparison with in-house connections, the price per m^3 paid at kiosks or pre-paid meters are often higher than lowest block tariffs for in-house connections. Schwartz et al., (2017) show that lower service levels provided through appropriate technologies are frequently more expensive than services provided through in-house connections and yard taps. They claim that, as a result, already existing social and economic inequities are reinforced through differentiated service provision.

A more general concern is that by emphasising payment for services, money becomes a prerequisite for accessing water. Access to water is cut off when consumers have used up their credit or when they do not have money to pay at water points (Harvey, 2005; Heymans et al., 2014). In those cases, they have either the option of trying to access potable water through their social network (neighbours, friends and family) or of accessing alternative water sources (rivers and lakes), which are frequently unfit for consumption. In cases where a consumer accesses water through their social network, it inevitably leads to a dependency of the consumer on his/her network.

2.3.3 Organisation: Community and Small-Scale Private Sector Involvement

As highlighted above, pro-poor water services often involve some replacement of money for in-kind contributions. This may happen by asking consumers to provide labour to construction and connection work. It often also happens by expecting a higher involvement of the users or the community in the actual management and operation of the infrastructure. This latter strategy makes it easier and cheaper for the utility to provide water to low-income areas, while also reducing the monetary costs of water for consumers. Two main narratives justify involving the community in providing water services. The first relates to the perception that a community-based organisation will more efficiently and effectively operate and manage a water supply system than a government utility. The narrative, which is often linked to neo-liberal calls for reducing state involvement in water services provision, argues that community members, being the end-beneficiaries of the water supply system, will have strong incentives to efficiently manage the water supply system. Moreover, by having a better understanding of their own needs, the community is also expected to better ensure that the system is managed in accordance with demands. Community involvement in this narrative is thus seen as contributing to sustainable water supply, on the condition that community members perceive it to be in their best interest to operate and manage a water system (Carter et al., 1999). The second framework argues that the delegation of management responsibilities is a form of empowerment because community members obtain a voice in how the system is operated and managed. This narrative thus emphasises the decision-making power granted to a community (Harvey and Reed, 2007).

Although community involvement may take on a number of different forms, in many cities in Sub-Saharan Africa, it de facto entails water utilities delegating the provision of services to community-based organisations or private organisations residing in the lower-income areas. In this arrangement, the water utility becomes a bulk supplier of water to these organisations. These organisations then both deliver the water to consumers and collect tariffs for this service. Examples include the management of water kiosks by water user associations (Rusca and Schwartz, 2012; Rusca et al., 2015) and the Delegated Management Model, which involves a partnership between a water utility and small-scale community-based private operators (Schwartz and Sanga, 2010). In the policy literature promoting these models, they are framed as reflecting a 'demand-oriented approach' as the organisational arrangement can be adapted to users' inclinations and provide a podium for the voices of users to be heard (Watson et al., 1997; Rusca et al., 2015). Studies documenting the implementation of these models, however, have raised a number of critiques and concerns. As early as 1980, community management was already viewed as a 'buzz phrase' and described as 'the mythology for the [Drinking Water] Decade' (Feachem, 1980: 15) that lasted from 1980 to 1990. Bell and Franceys (1995: 1174) argued that except for small projects being based on 'personal loyalty, civic duty and a cooperative spirit', community participation is often challenging and does little to lower costs. Gomez and Nakat (2002) even suggest that community involvement can lead to higher costs. Studies also show that empowerment, one of the underlying aims of community involvement, has, in practice, largely been stripped of its radical and transforming qualities and has become individualised (Cleaver, 1999). As communities are characterised by their own internal differences and politics, community management can lead to the reproduction of existing inequalities within these communities (Manor, 2004: 192). Community involvement may actually disempower local citizens as some are excluded from meaningful participation (Dill, 2009).

Another concern raised by the more critical literature relates to the requirements for sustainable management of systems. Without support by governments or non-governmental organisations (NGOs), the ability and motivation of communities to keep water systems running and keep collecting revenues for recurring expenditures may decline (Carter et al., 1999: 295). Similarly, small-scale private operators may not have the required capacity to operate and manage a water system (Schwartz et al., 2017; Tutusaus et al., 2018). In actual practice, management through community-based organisations or small-scale providers may imply that the State withdraws from its responsibility of service provisioning and fails to provide such support (Chapter 4; Rusca et al., 2015; Schwartz et al., 2017). This is an important conclusion of Frediani's (2015) investigation of small-scale private water providers operating under the Delegated Management Model. He finds that risks and responsibilities are transferred to small-scale private operators, who lack the capacity to adequately deal with these responsibilities and who are not sufficiently supported in their operations. This leads to a lower level of service provided to customers served by these intermediaries.

A final criticism relates to the need of the intermediary organisations to operate on the basis of cost recovery. They often receive no subsidies from the government. Chapter 4 and Rusca, and Schwartz (2017), for instance, found that community-based and private operators both operate like business entities that need to recover their investment costs and make a profit. As a result, water services provided through such intermediaries may turn out to be more expensive than what users would have paid when paying the tariffs charged for water delivered through a connection to the conventional network. In the case of Lilongwe, Malawi, for instance, the water user associations triple the (subsidised) price paid for bulk water when charging users of the water kiosk. As a result, water kiosk users pay 1.5 times more for their water than users connected to the conventional network (Rusca and Schwartz, 2017).

2.4 POPULARITY OF PRO-POOR APPROACHES: BALANCING FINANCIAL AND SOCIAL OBJECTIVES

Despite the criticism of pro-poor water service approaches, they remain popular as approaches to improve service provisioning to low-income areas. In our analysis, an important explanation for this is that they fit the mixed mandate that public water utilities have in developing countries: public water utilities have to satisfy both social and commercial objectives. On the one hand, an increasing consensus appears to exist that public water utilities should operate on the basis of cost recovery. Pricing policies, which fail to recover costs, are considered unsound as they weaken the financial viability of the utility (Kessides, 2004). At the same time, international and national treaties and policies emphasise that access to water is a human right to underscore that it should be (made) available to all. In fact, most countries have committed themselves to the achievement of Goal 6 of the Sustainable Development Goals (SDGs), which states that sustainable management of water and sanitation for all is to be achieved by 2030. Realising this ambitious goal largely falls on water utilities. To be able to achieve it without compromising their financial or commercial viability, what has come to be known as pro-poor services provides an ideal approach (WSP, 2008: 12).

2.5 CONCLUSION

The concept of pro-poor services is very popular in the water supply and sanitation sector. The policy literature on the topic promotes a pro-poor services approach by emphasising how it can help improve access for poor consumers living in low-income areas. Unlike the MII, which advocated for a standardised level of service for all, pro-poor services entail an explicit differentiation of services by distinguishing between areas and consumers on the basis of income, land tenure or ability to pay. This differentiation happens by adapting the technology to the abilities and needs of low-income users and by the design of specific financial and organisational arrangements. As the name 'pro-poor services' suggests, pro-poor services are presented as being, first and foremost, beneficial for the poor. Our analysis shows that 'pro-poor services' are also an attractive way for water utilities to fulfil their dual objectives of ensuring commercial viability, while expanding services to the poor. In contrast to a 'one-size-fit-all' approach, pro-poor services justify and produce a differentiation, which enables public water utilities to delegate some of the costs, burdens and risks of providing water to low-income areas to the inhabitants of these areas. It allows them to maintain credibility in the eyes of those donors and governments demanding that water services should be available to all, without compromising their financial sustainability. We conclude that the different dimensions of pro-poor service provisioning work out well for water utilities: the use of low-cost technologies, limiting the consumption of consumers in low-income areas, the guaranteed payment for services and the use of distinct organisational structures makes it possible for the water utility to meet social objectives, while continuing to focus most investments and efforts on serving existing 'normal' customers.

This dual benefit of supplying the poor while ensuring commercial viability is frequently presented as a win–win situation: both the users in low-income neighbourhoods as well as the water utilities providing such services are believed and said to gain. With water utilities operating on commercial principles, the differentiation of service levels is presented as 'a medium for progressive steps in favour of the poor and a driving force behind the necessary regulation of composite supply systems' (Jaglin, 2008: 1898). While pro-poor services may indeed increase or improve the provision to and accessibility of water in low-income areas, our review and analysis show that it risks widening existing inequities in water access: the urban poor may have access to poorer services for which they pay more. If this happens, pro-poor strategies become against-poor and the assumed win–win scenario becomes a win–lose one. Unless the implementation of these strategies is linked to explicit concerns about the quantity, quality and affordability of the service to the poor, pro-poor services will not lead to equitable access.

3

ALIGNING STAKEHOLDER INTERESTS: HOW 'APPROPRIATE TECHNOLOGIES' HAVE BECOME THE ACCEPTED WATER INFRASTRUCTURE SOLUTIONS FOR LOW-INCOME AREAS IN KENYA[3]

Service differentiation and the use of 'appropriate technologies' are often presented as alternative approaches to providing water service to low-income areas. Service differentiation allows water service providers to manage the risks associated with service expansion to these areas. We argue that rather than a choice of the service provider alone, the decision for service differentiation should be seen as a consensus between different actors in the water services sector. A review of secondary data, and semi-structured interviews with diverse stakeholders in three Kenyan cities, show that managers of water utilities also choose these strategies because they appeal to and align with the interest of donors, governments, sector organizations, and alternative providers.

---

[3] This chapter is based on: Boakye-Ansah, A. S., Schwartz, K., and Zwarteveen, M. (2020). Aligning stakeholder interests: How 'appropriate' technologies have become the accepted water infrastructure solutions for low-income areas. Utilities Policy, 66, 101081.

3.1 INTRODUCTION

Globally, poor access to safe drinking water sources remains a challenge. In spite of the substantial progress made in the past decades towards mitigating this challenge, about two billion people, mostly in developing countries, still do not have access to safely managed drinking water sources (WHO and UNICEF, 2017). Access to such improved water sources is especially limited for households in low-income areas in which a large portion of the urban population resides (Kjellen, 2000; Sansom and Bos, 2008). In Kenya, for example, while 84% of households in high and middle-income settlements in the cities have access to water through piped sources, only about 34% of households in low-income areas within these cities are served through such water sources (World Bank, 2015). In an effort to address such supply inequalities in these cities, water utilities are increasingly employing service differentiation as their approach to service provision. Service differentiation "entails explicitly distinguishing between different water service strategies and between different customers or clients" (Chapter 2:12)[4]. This differentiation is visible both in the technologies used to provide water services and in the pricing regime that is attached to the service provided. The technology may vary from in-house connections for water supply through so-called 'appropriate technologies', such as water kiosks, where water is sold in 20-litre jerry cans (Chapter 4). Pricing regimes range from those used for in-house connections, which usually follow some variation of block-tariffs, to kiosks, which charge consumers for each jerry can, filled at the kiosk. The argument in favour of service differentiation is that it allows a water provider to tailor the level of service provided to the demands and capacities of consumers (Mara and Alabaster, 2008). It is precisely this matching of consumers with a particular technology and pricing regime that makes a technology 'appropriate'[5] . The underlying idea is that poor consumers, who have relatively little disposable and often variable income, can afford a different level of service (and associated pricing regime) than middle- and high-income households. Although service differentiation entails distinguishing between different consumer categories within a city, it also reflects, "a pragmatic move towards accommodating social and spatial disparities" (Jaglin, 2008: 1898).

---

[4] In water services management and governance literature, the concept of service differentiation has been critiqued from an equity point of view. Some critical studies suggest that the use of service differentiation leads to the poor having to pay more for their water services than the better-off. Our aim in this article is not to argue in this line or critique service differentiation but to show how it is used by the different stakeholders in the water sector to advance their interests. For more literature on service differentiation and equity in access see for instance (Abubakar, 2019; Bayliss and Fine, 2007; Bond, 2004; Goff and Crow, 2014; Hukka and Katko, 2003; McDonald and Ruiters, 2005; Pihljak et al., 2019; Rusca and Schwartz, 2018; Schwartz et al., 2017; Wilder and Romero Lankao, 2006).

[5] The term 'appropriate technologies' was popularised in the 1990s when it replaced the term 'low-cost technologies'.

The strategy for pursuing service differentiation is usually presented as a conscious decision by the management of the water utility (Berg and Mugisha, 2010; Schwartz et al., 2017) to deal with the challenges and risks associated with service expansion to low-income areas. These include the low water demand of poor consumers; the challenge of collecting bills in these areas; and a proliferation of illegal connections, which increases levels of non-revenue water (Chapter 2; Heymans et al., 2014). Service differentiation reduces the risks of providing services to low-income areas (Chapter 2; Schwartz et al., 2017) because extending services through 'appropriate technologies' significantly reduces the investment cost per person served. In comparison with in-house connections, which may serve only about six persons, a single water kiosk can serve up to 200 persons (Chapter 4). Moreover, most appropriate technologies adhere to a form of pre-payment, where the consumer pays prior to consuming water. This ensures that the utility does not face the challenges of collecting water bills.

While acknowledging the link between service differentiation and the reduction of risks faced by water utilities (Chapters 2 and 4), this article argues that the proliferation of 'appropriate technologies' as a strategy of servicing low-income areas (LIAs) is not just a risk-reduction strategy of utility managers. Rather, the popularity of the 'appropriate technologies' approach to service expansion in low-income areas also stems from its ability to align the interests of a range of stakeholders in the water services sector. Hence, the use of appropriate technologies is something that appeals not just to utility managers, but also to donors, governments, sector organizations, and alternative providers. The choice for appropriate technologies is therefore the outcome of aligning the preferences and interests of a diverse group of key stakeholders. This argument is developed through a case study of three Kenyan water utilities (Water Service Providers-WSPs), which have implemented water service interventions in the low-income areas within their services areas. In Kenya, provision of water services is mandated to WSPs that are owned by county governments but operate as autonomous utilities. The main argument of our study is developed by first describing the case studies and the interventions these utilities implemented in low-income areas. We then examine how different stakeholders benefit from the type of interventions employed by the WSPs.

3.2 METHODOLOGY

This study formed part of a larger study carried out in Kenya to investigate how water utilities address the challenge of providing equal services within the fragmented cities they operate (Chapter 4, Kemendi and Tutusaus, 2018; Schwartz et al., 2017). Field research for the development of these cases was carried out in four phases: January to April 2017, November to December 2017, from January to April 2018 and from May to June 2019. For each phase, a mixed-method approach was employed. A mixed-method approach was used as it provided the opportunity to triangulate and crosscheck data obtained from both primary and secondary sources. Secondary quantitative data was obtained from the water utilities for a period of five years (2013-2017). This secondary data concerned reports on 1) water supply interventions implemented in the low-income areas, 2) revenue collection rates associated with the service modalities, 3) non-revenue water rates associated with the modalities, and 4) the extent of service coverage associated with the modalities. This was followed up with the collection of qualitative data through semi-structured interviews with open-ended questions with 80 key respondents. The key respondents were selected through snowball sampling. The low-income area coordinators of the water utilities studied for this article identified initial contacts. The key respondents included managers of the water utilities, alternative water providers and operators of water points in the LIAs, officials of governmental agencies (including staff of the Water Services Regulatory Board (WASREB), the Water Sector Trust Fund (WSTF), Ministry of Water and Irrigation-MWI, County Water Directorate and the Water Services Board-WSB). It also included international NGOs and donor organisations (including the World Bank, Deutsche Gesellschaft für Internationale Zusammenarbeit-GIZ, VEI Dutch Water Operators, Water and Sanitation for the Urban Poor –WSUP, and SNV Netherlands Development Organization) and local NGOs (including Upande project, Umande trust, SANA international, Kenya Water and Sanitation Civil Society Network-KEWASNET, Kisumu Urban Apostolate Programs-KUAP and Water Action Groups-WAGs). The interviews which lasted for an average 30 minutes focused on 1) the institutional arrangement under which the water utilities operate, 2) financial flows related to service provision in low-income areas, and 3) decisions relating to the selection of particular water supply strategies. Respondents agreed verbally to partake in the study after they had been duly informed of the purpose and aims of the study. While participants in this study were not anonymous, all identifying information linking them to the study was kept confidential. This meant presenting their views without referencing their identities. Relevant data from notes taken during the interviews were coded (Table 1). The data were analysed through a grounded theory approach with constant comparison of data collected from the different respondents with the secondary reports. This process helped bring up additional questions and subsequently aided in further data collection until a point of saturation, where a conceptual framework that is a practical representation of the research was developed.

Table 1: Coding of respondents and their characterization

Code	Respondent/organisation code is applied to	Number of people interviewed
KWSP-NAWASSCO	Nakuru Water and Sanitation Company	10
KWSP-KIWASCO	Kisumu Water and Sanitation Company	15
KWSP-KEWASCO	Kericho Water and Sanitation Company	8

Code	Respondent/organisation code is applied to	Number of people interviewed
KGA-WASREB	Water Services Regulatory Board (WASREB)	5
KGA-WSTF	Water Sector Trust Fund (WSTF)	5
KGA-MWI	Ministry of Water and Irrigation (MWI)	4
INGO	International Non-governmental Organisations	8
LNGO	Local Non-governmental Organisations	10
AWSP/O	Alternative Water Service Providers/Water point Operators	15
Total Number of Stakeholders interviewed		80

Source: Author.

3.3 CASE STUDY

The cases selected for this study are Kisumu Water and Sanitation Company (KIWASCO), Nakuru Water and Sanitation Company (NAWASSCO), and Kericho Water and Sanitation Company (KEWASCO) (Figure 2). These cases were selected because all three water utilities have established interventions through which they supply water to the low-income areas within their service areas. The utilities have adopted different types of approaches and technologies for these interventions as discussed below (Table 2).

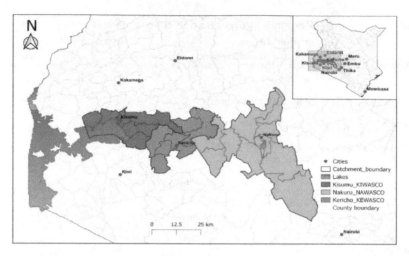

Figure 2: map of Kenya showing the location of the WSPs.
Source: Author.

Table 2: Characteristics of the three Water Service Providers

Indicator	KIWASCO	NAWASSCO	KEWASCO
Total population in service area	449,012	510,791	188,577
Total population served	342,203	458,965	100,956
Active connections	54,989	51,396	13,362
Staff	331	219	133
Non-revenue water (%)	37	36	51
Total water produced ('000 m^3)/year	9,455	12,655	3,659
Turnover (US$)/year[6]	6,594,040	8,144,480	1,699,880

Source: Author; adapted from data collected from the three WSPs.

KIWASCO, a water utility that began operations in 2003, is mandated to supply water to the inhabitants of Kisumu city (Owuor and Foeken, 2012). About 60% of the approximately 450,000 people within their service area live in low-income areas (Water Services regulatory Board (WASREB, 2019). Meanwhile, the company's operations cover about 76% of the entire population within its service area (WASREB, 2019). The

---

[6] KES 1 = US$ 0.00934.

main intervention in low-income areas used by KIWASCO is the Delegated Management Model (DMM). Developed in 2004, through a partnership between KIWASCO, the Water and Sanitation Programme-Africa (WSP-Africa), and the French embassy, the DMM involves KIWASCO supplying water in bulk to small-scale operators (individuals or groups), called master operators (MOs). The MOs then distribute the water to consumers, either through on-premise connections or through water kiosks and yard taps. They are also responsible for collecting the bills associated with water supply in the low-income areas and paying KIWASCO for the bulk water supplied. 39 of such DMMs currently exist in six out of the 22 low-income areas in Kisumu. With the DMM, the issue of land tenure as a requirement for connecting to utility water is eliminated: residents who have the means, pay for individual on-premise connections. Besides the DMM approach, KIWASCO also uses water kiosks to supply water in areas where the DMM has not been implemented. A total of 340 water kiosks exist in 16 low-income areas. These water kiosks are operated with a post-paid meter (295) or with a pre-paid meter (45).

NAWASSCO has a mandate to supply water to the approximately 510,000 inhabitants of the city of Nakuru (NAWASSCO, 2013). About 40% of the city's population lives in low-income informal areas (Chakava and Franceys, 2016). NAWASSCO's operations cover 90% of the population within its service area (WASREB, 2019). The interventions employed by NAWASSCO to supply water to the LIAs in Nakuru include water kiosks and yard taps (with pre-paid or post-paid meters). Of these interventions, the pre-paid metering (ppms) system is the preferred option[7]. The technology involves the use of magnetic or radio frequency identification (RFID) tokens that customers use to purchase water from communal points/yard taps. Between the periods of 2012 and 2016, the company installed 260 pre-paid meter taps at various locations within the low-income areas[8]. In addition to the prepaid meters, a number of water kiosks (40) and yard taps with post-paid meters (112) are also used by NAWASSCO to supply water in the LIAs.

 KEWASCO, a company that started operations in 2002 is mandated to supply water to about 180,000 people within Kericho town and the urban centres of Kapsoit, Kapsuser, and Brooke (KEWASCO, 2018). About 34% of the population the company serves live in low-income areas. KEWASCO's operations cover about 54% of the population it is mandated to serve (WASREB, 2019). As for Kericho, the WSP (KEWSCO) prefers to supply water to the LIAs through water kiosks[9]. Subsequently, 23 water kiosks were put up in different locations within the LIAs. Meanwhile, reasons ranging from non-payment

---

[7] Interview KWSP-NAWASSCO, 15 February, 2017.

[8] The system of pre-paid dispensers, however, has faced several challenges, leading to the failure of some meters. At the time of the fieldwork for this study 30 (12%) out of the total number installed were operational.

[9] Interview KWSP-KEWASCO, 27 March, 2017.

of bills by operators to technical mishaps rendered most of these kiosks non-operational. At the time of this study, only six were operational. Besides the kiosks, the yard tap is another intervention employed by KEWSCO, eight of which are located within different households in the city.

3.4 RESULTS: UNPACKING THE INTERESTS AND PREFERENCES OF THE DIFFERENT ACTORS

As indicated in the previous section, a number of water services strategies have been implemented in Kenya's fast-growing cities to meet the water demands of low-income communities. In this section, we unpack the interests and preferences of the different actors in the choice of these interventions for meeting the water needs of consumers in low-income areas.

3.4.1 The Kenyan Government

National statistics indicate that only about 20 percent of Kenyans living in urban low-income areas have access to safe water. Not surprisingly, the extension of services to these areas figures prominently in various laws and regulations emanated by the Kenyan government. The Kenyan constitution declares, "Every person in Kenya has the right to clean and safe water in adequate quantities" (Government of Kenya (GoK), 2010:31). The Kenya Vision 2030 stipulates the achievement of universal service coverage, reflecting the government's commitment to achieving target 6.1 of the Sustainable Development Goals, which stipulates that universal service coverage is to be achieved by 2030. In response to these objectives, the Kenyan government, through the Ministry of Water and Irrigation (MWI) adopted the National Water Services Strategy (NWSS) with a particular focus on servicing low-income areas. An important element of this strategy focuses on designing and enforcing national standards for low-cost technologies (GoK, 2007a). In a move to operationalise this strategy, the MWI adopted the Pro-poor Implementation Plan for Water Supply and Sanitation (PPIP-WSS) (GoK, 2007b). The PPIP-WSS spells out the actions that need to be carried out by the different actors and institutions in order to extend water supply to low-income areas. More specifically, the PPIP-WSS indicates the intention to reach the poor at a rate of one million people per year.

Yet, the actual budget available for investment in the expansion of water services is limited. The National Water Master Plan (NWMP) estimates the total cost of investment and rehabilitation needed for water supply until 2030 to be about US$20 billion[10]

---

[10] Ksh 1.7 trillion (NWMP, 2013).

(NWMP, 2013). Without this investment, the goal of universal access to water services cannot be achieved by 2030. According to that same document, however, the available government budget for water supply is only about US$7 billion, leaving a shortfall of approximately US$ 13 billion for the water services sector in Kenya (WASREB, 2016)[11]. Another report suggests that on an annual basis US $727 million is to be invested for the goals of universal service coverage to be achieved; meanwhile, only a third of this amount is actually invested annually[12] (WASREB, 2015). Of the funds invested in the Kenyan water services sector, the overwhelming majority consists of donor funding. It was estimated that during the fiscal year 2013-2014, 94% of the investment funds were provided by donors (WASREB, 2014:9). The Government of Kenya is thus confronted with a significant funding gap and with a high level of donor dependency. Given that donor funding in the Kenyan water sector will decline in the coming years (Savelli and Schwartz, 2019), the funding gap for the government of Kenya will most likely increase in the coming years.

The significant funding gap means that the government of Kenya is strongly dependent on identifying low-cost options (including technologies) for achieving the ambitious policy targets of universal service coverage stipulated in the Kenyan Constitution and Kenya Vision 2030.

3.4.2 The Donors

In Kenya, donor involvement in the water sector was institutionalised by the Water Act 2002 (section 83, 3c) and has been reinforced under the current 2016 Water Act (sections 117d and 126c). Funding for the extension of water services can take different routes. One funding arrangement is to channel funding through the Water Sector Trust Fund (WSTF). The WSTF is a Kenyan water sector organization, established through the Water Act 2002, which is mandated to finance water and sanitation services for the poor and underserved communities. About 80% of the funds of the WSTF come from multilateral as well as bilateral donors (Table 3).

---

[11] Ksh 592.4 billion (NWMP, 2013). The figures mentioned in the National Water Master Plan assume an exchange rate of Ksh 85.24 = US$ 1.00. The current exchange rate, however, is approximately Ksh 100 = US$ 0.94 (June, 2020).

[12] Exact estimates of the required investment vary. Although, WASREB (2015) suggests a figure of US$727 million other organizations, like KIFFWA, funding up to US$1.9 billion is required annually (http://www.kiffwa.com; accessed 23 October, 2018).

Table 3: Summary of funding to the Water Sector Trust Fund (WSTF) for the period 2009-2017

Funding Partner	Support Amount (KSH)	% of Support
Government of Kenya (GoK)	2,173,592,097.00	**20**
Donor organization		**80**
Kredit fuer Wideraufbau (KfW)	2,498,597,783.71	23
World Bank	198,650,640.95	2
African Development Bank (ADB)	50,348,530.50	0.5
European Union (EU)	1,260,931,819.49	11
Deutsche Gesellschaft für Internationale Zusammenarbeit (GIZ)	10,522,216.40	0.1
Bill and Melinda Gates foundation	299,653,965.78	3
Swedish International Development Cooperation Agency	1,242,086,372.80	11
Danish International Development Agency (DANIDA)	254,679,244.00	2
United Nations Human Settlements Programme (UN-HABITAT)	38,983,890.65	0.4
United Nations Children's Fund (UNICEF)	1,264,168,144.10	12
Government of France (GoF)	621,232,384.20	6
Embassy of Israel	172,000.00	0.002
Medium-Term Arid and Semi-Arid (ASAL) Programme (MTAP)	909,000,000.00	8
International Fund for Agricultural Development (IFAD)	147,402,890.00	1

Source: Author; adapted from data collected from the Water Sector Trust Fund Kenya.

Almost all donor funds channelled through the WSTF are full grants, except that of KFW and the World Bank, which have components of their support implemented through Aid on Delivery and Output-based Aid (WSTF, 2016)[13]. In addition to donor funds channelled

---

[13] These are grants provided as subsidies to water utilities that access commercial loans for investment purposes. Utilities are pre-financed by loans from commercial domestic lenders on market terms and then the AoD and OBA.

through the WSTF, some donors also provide funding for water projects directly to the WSPs or through local and sometimes some international NGOs. These direct funds also come in the form of grants, loans payable with no interest, or co-funding[14] .

Donors are organisations that have been created with the aim of achieving certain policy targets or objectives of the funders they represent. Frequently, these policy objectives are linked to global agendas such as the Sustainable Development Goals. For example, the WASH policy of the Dutch Ministry of Foreign Affairs stipulates in its WASH Strategy 2016-2030 that the Dutch government has committed itself "to provide 50 million people with access to sustainable sanitation and 30 million people access to sustainable drinking water by 2030" (Ministry of Foreign Affairs of the Netherlands, 2016). For employees of donor agencies, achieving these targets is important.

For these organizations, the use of 'appropriate technologies' is a way to ensure that they will be able to make substantial gains in service coverage quickly. For instance, the funds from GIZ to the WSTF for pro-poor activities support the construction of water kiosks. Each kiosk is estimated to serve at least 200 people. By supporting a number of these kiosks in about seventy (70) water utilities, the GIZ reports improvements in water services access for some 1.6 million people. Also, in Nakuru, the American donor USAID insisted that the pre-paid water dispensers (PPDs) they financed for water provisioning serve 50 households, even though these dispensers were not really designed for such use[15] . However, placing them in compounds with more users allows the donor to claim a more significant contribution to improved access to water services. This same logic is echoed by a representative of the Dutch Ministry of Foreign Affairs who explains that "[w]e know that small and slow often works better, but donors are under pressure to produce large and fast" (Savelli and Schwartz, 2019). Indeed, 'appropriate technologies' are favoured by donors precisely because they allow them to produce 'large and fast' with limited investments.

Grants are used to subsidize a certain percentage of the cost of the loans. While KFW provides KFW 50% subsidy, the World Bank provides a 60% (WASREB, 2016).

[14] Here the donor provides a percentage of the funds needed for a project. Actual percentages depend on the particular donor organization providing the support. However, at all times, the WSPs are required to show proof of their ability to provide the remaining percentage.

[15] Interview, KWSP-NAWASSCO, 2 May, 2019. The interviewee also indicated that the frequent use of these dispensers led to frequent breakdowns of the dispenser.

3.4.3 The Water Service Provider

As highlighted earlier, the Kenyan government has repeatedly stated that it pursues a goal of ensuring universal access to water services by 2030. The Water Service Providers (WSPs) are the main structures to achieve this policy goal. This means that the WSPs have to show that they are contributing to achieving universal access. At the same time, the enactment of the Water Act 2002 saw the introduction of utility reforms into the water sector[16]. An important focus of these reforms was that water utilities were to become commercially viable entities. The Act requires these WSPs to operate as autonomous entities "…..providing the water services … on a commercial basis and in accordance with sound business principles" (Water Act, Section 57(5d)). In line with the principles of commercialization, the "… tariffs were set to cover full cost recovery in urban schemes and operation and maintenance cost in rural schemes" (GoK, 2002: 87). In line with these demands, the benchmarks and indicators set by the Kenyan regulator (WASREB) to assess the performance of the WSPs have a strong focus on cost recovery[17].

When the dual objectives of universal service coverage and cost-recovery are combined, the expansion of services to low-income areas appears as a "capital-intensive high-risk venture in commercially non-viable spots"[18] – simply put a "risky business". In developing service provision to low-income areas, WSPs search for approaches or strategies that reduce these risks. Service differentiation, in which 'appropriate technologies' play a prominent role in servicing low-income areas, is such a risk-reducing approach for the WSP. The use of 'appropriate technologies', as a manager of one of the WSPs described it, represents "the best fit for water supply in the low-income areas"[19], at least from the WSP perspective.

In addition to challenges emanating from combining water service provision to the LIAs with cost-recovery objectives, the tariff structure handed down to the WSPs by the regulator (WASREB) also pushes them to adopt appropriate technologies. This tariff structure places WSPs into one of three categories as illustrated in Table 4.

---

[16] Efforts towards restructuring and commercialising water supply began in 1987 by the government, through the Urban Water and Sanitation Management Project (UWASAM), which led to the establishment of the Water and Sewerage Operations Unit (WSOU). This project was initially piloted (1987-1993) in three municipalities (Kitale, Kericho, and Nyahururu) and was subsequently extended (1994-1996) to include five more municipalities (Kisumu, Eldoret, Nakuru, Nanyuki, and Thika) (Onjala, 2002).

[17] WASREB impact reports, Issues 1-11.

[18] Interview, KGA-WASREB, 10 February, 2018.

[19] Interview, KGA-KEWSASCO, 6 March, 2017.

Table 4: Classification of WSPs with respect to tariff structure

Utility classification	Indicator	Remarks
Type 1	WSPs cannot meet its full coverage of operations and maintenance (O&M) cost	WSPs in this category operates with deficit and mainly rely on government subsidies to maintain its operations.
Type 2	WSPs that can meet its (O&M) cost but lacks the ability to pay its debts.	For WSPs in this category, tariffs are structured to meet O&M costs. They do not have the capacity or surpluses to invest in infrastructure/other investments
Type 3	O&M costs are covered between 100% and 150% and repayment of debts is achieved	For WSPs in this category, tariffs are structured to meet O&M cost, with a surplus to carry out investments.

Source: Author; adapted from data collected from the Water Services Regulatory Board WASREB.

The three WSPs in this study all fall under the type 2 classification. As such, the revenue they accrue from tariffs cannot be channelled into investments in infrastructure for service extension. To extend or improve water supply in the low-income areas, these WSPs are thus completely dependent on the WSTF or donors as a way of "securing funds for investing in water supply in low-income areas within their service areas" [20] (see also Table 5).

The underlying idea is that such funding provides WSPs with the "opportunities to leverage their commercial viability against the intended risk of expanding services to these areas"[21] . By using external funding, WSPs do not only expand service coverage in the low-income areas, but the overall coverage within their service areas also improves (Table 6). As 'appropriate technologies are frequently associated with low water losses and high billing/collection ratios, both revenue collection and non-revenue water (NRW) rates associated with these interventions improve (Figure 3 and Figure 4).

---

[20] Interview: KGA-WSTF, 3 February, 2017.

[21] Interview KWSP-WSTF, 6 February, 2018.

Table 5: Summary of projects carried out by the WSPs and their source of funding (2013-2017)

WSP	Number of projects	Project Target		Utility Funded	Donor funded	
		LIAs	Others	Utility Funded	WSTF	Others
KEWASCO	5	3	2	0	3	2
NAWASSCO	8	8	0	0	4	4
KIWASCO	9	10	1	1	2	8

Source: Author; adapted from data collected from the three WSPs.

Table 6: Increased water coverage with the implementation of interventions in LIAs (2013-2017)

Intervention	Kisumu		Nakuru		Kericho	
	Number	Population covered	Number	Population covered	Number	Population covered
Water kiosk						
with post-paid meters	295	59000	40	21500	6	1800
with prepaid meters	45	9000	0	0	0	0
Yard taps						
with post-paid meters	0	0	177	5850	8	1600
with pre-paid meters	0	0	83	19920	0	0
DMM	39	152800	0	0	0	0
Total		220800		66000		3400
% pop covered in low-income areas		74		22		5
% pop covered relative to pop in service area		44		10		2

Source Author; adapted from data collected from the three WSPs.

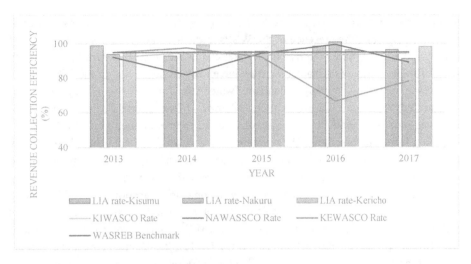

Figure 3: revenue collection efficiency for (a) KIWASCO (b) NAWASSCO (c) KEWASCO.
Source: Author.

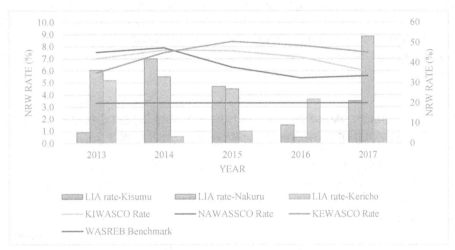

Figure 4: NRW rate from pro-poor intervention in (a) KIWASCO (b) NAWASSCO (c) KEWASCO.
Source: Author.

3.4.4 The Small-Scale Provider (The Cartels)

Other prominent actors in water service provision in LIAs are informal private water providers. The inability of formal WSPs to supply water to low-income areas in Kenyan cities led to a proliferation of such informal private providers. These informal providers supplied water from boreholes and wells or from (illegal) connections to utility networks. For these providers, "the presence of the WSPs in the low-income areas meant a loss of their investments"[22] . Often referred to as water cartels due to the uncontrolled and rowdy nature of their operations (Chapter 4), they have resisted operations of the WSPs in the low-income areas they serve by vandalising water infrastructure, connecting illegally to water sources and controlling water prices. Members of these cartels, many of whom are also owners of the lands in these areas, often refuse to give up their lands for the WSPs to develop infrastructure. When they did agree to give up their lands, they demand very high prices. For WSPs the use of 'appropriate technologies' represents an opportunity to form partnerships with these operators. To the WSPs, the use of "pro-poor interventions provided them with the means to enter the low-income areas, absorb the cartels into their formal operations so that they can be able to control their activities and, in the end, control the illegalities they perpetrate in the low-income areas"[23] . Such partnerships were formed with Master Operators (MOs) who operated the DMMs, kiosks operators, and landlords on whose lands yard taps were located (Chapter 4). With such partnerships, the risks for the WSPs are limited; they supply bulk water at fixed rates to the small-scale operators, while they can still claim to contribute to achieving universal coverage. This is why the WSPs consider such partnerships as a win-win situation: "they get to extend water to the low-income areas while the cartels are given the official recognition to operate"[24] . At the same time, the cartels benefit from such an approach to servicing low-income areas as they get (legal) access to water at subsidized bulk rates from the WSPs and retain the consumer market they serve in these areas.

3.5 Discussion: Aligning the Interests and Interdependencies of Stakeholders and Appropriate Technologies

In much of the literature on service differentiation, the decision to extend services to low-income areas through appropriate technologies is largely portrayed as a decision of the utility to better match the demands of consumers (Berg and Mughisha, 2010; Cross and Morel, 2005; Mara and Alabaster, 2008; Heymans et al., 2014; Schwartz et al., 2017). While acknowledging that the use of appropriate technologies allows the utility to extend services to these areas while minimising the commercial risks, the decision to use such technologies is not just a choice by the managers of the water utility. Rather, the decision to opt for such technologies is the result of a consensus negotiated between different

---

[22] Interview, AWSP/0, 15 February, 2017.

[23] Interview, KWSP-KEWASCO, 17 March, 2017.

[24] Interview, KWSP-KIWASCO, 18 February, 2018.

stakeholders, who are dependent upon each other. In this sense, appropriate technologies represent a way to align the interests and preferences of these diverse stakeholders (Figure 5).

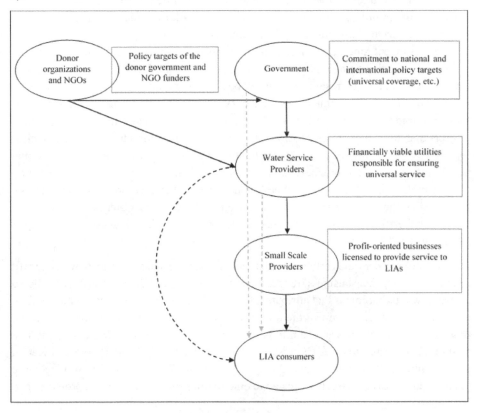

Figure 5: framework showing how stakeholders and their interests are aligned. Source: Author.

Ideal water utilities are typically depicted as autonomous business-like entities that provide a basic service (Baeitti et al., 2006). Yet, the everyday reality of most water utilities in developing countries is that they operate in a complex institutional environment involving multiple stakeholders. What characterizes this environment, is that the different stakeholders are dependent on each other for meeting their needs and satisfying their interests. Water utilities are subject to the laws, policies, and regulations emanating from the county and national governments. At the same time, they depend on the political support from politicians elected to represent these governments. These utilities also crucially depend on the donors, whose help is needed to fund service extensions to low-income areas. By opting for service differentiation and associated appropriate technologies, utility managers can navigate, align, and comply with these diverse demands and pressures.

For the Kenyan national government, which has committed itself to universal service coverage, water utilities are its main vehicle or tool to achieve this commitment. Moreover, given the large funding gap, it requires donors and NGOs to contribute to the needed investment funding. With funding from donor organisations, the government can claim to pursue this ambitious goal, without having to significantly raise investment in water supply from its own sources in order to cover the funding gap. These donors, who often have linked development goals such as extending services to the poor to their economic trade agenda (Savelli et al., 2018), also need the water utilities to implement projects that allow them to demonstrate that they are contributing to achieving universal service coverage. They often use quantifiable targets such as the number of additional people that have received access to water through their investments to convince their constituencies that they spent their money well. Funding low-cost appropriate technologies allows high numbers of additional people accessing water at a relatively low cost. It is therefore a favoured strategy of donor organisations and NGOs, making it relatively easy to claim success.

Small-scale providers and cartels that are involved in the final distribution of water benefit from maintaining their business. Rather than operating under a continuous threat of being closed down, technologies that attribute a formal role to small-scale operators in the final distribution of water legitimize the operations of these providers. By using small scale providers, water utilities can claim to both address social objectives while doing so in a commercially viable manner (Bayliss and Fine, 2007; Bond, 2004; Hukka and Katko, 2003; Jaglin, 2002; World Bank, 2000; Younger, 2007). The risks of serving these 'challenging' areas are largely allocated to other parties such as small-scale providers and consumers.

3.6 CONCLUSION

Service differentiation has become the accepted approach to extend services to low-income areas. The choice for service differentiation and the associated use of appropriate technologies is, however, not only a decision of the water utility, but rather the result of a consensus or compromise among different actors that constitute the broader institutional environment in which the water utility operates. In the three cases studied in this article, appropriate technologies and service differentiation appeared to best satisfy the diverse preferences and interests of these actors. What these findings suggest is that understanding decisions about service differentiation and the use of particular technologies to service low-income areas is not simply a matter of engaging with the water utility, but requires engagement with the broader institutional environment in which the service provider operates.

Although the way in which services are provided to low-income areas is the result of an interaction between different actors, one important stakeholder appears to be missing from this decision-making process. How the consumer in low-income areas is involved in decision-making is unclear. In pursuing appropriate technologies, an apparent paradox then becomes visible. On the one hand, these technologies are advocated because they are claimed to best match the demands and capacities of a specific category of consumers, that is, those in low-income areas. This claim suggests the inhabitants of low-income areas 'demand' these technologies and the levels of service that they provide. The term 'appropriate' suggests that these technologies are most suited to the inhabitants of low-income areas. At the same time, there is very little research on service differentiation, its actual implementation, and how it is perceived and experienced by consumers. This raises questions about the actual 'demand' for appropriate technologies by these inhabitants. One of the few studies conducted on water service provision in Kenya suggests that consumers were very dissatisfied with the level of service provided by certain appropriate technologies such as water kiosks (Gulyani et al., 2005) and may well have very different ideas about the most suitable way of servicing low-income areas. If consumers are dissatisfied with appropriate technologies, they are not well-served and the alignment of producer and other stakeholder interests in the water services sectors occurs at the expense of consumer interest.

4

FROM ROWDY CARTELS TO ORGANIZED ONES? THE TRANSFER OF POWER IN URBAN WATER SUPPLY IN KENYA[25]

Due to the limited presence of formal water utilities in urban low-income areas in many developing countries, water supply is carried out by informal private providers. As part of efforts towards improving water services to low-income areas, the activities of these informal providers are currently being acknowledged through various formalization approaches. This article investigates one such formalization approach, which consists of partnerships between the utility and the informal providers. This is an approach which allows the utility to partially withdraw from service provisioning to the low-income areas. Based on empirical evidence from three such partnerships in low-income areas in three Kenyan cities, we show in this article that formalization redefines and strengthens the legal capacity of informal providers to gain control of service provision in these areas. This can bring benefits for the utility as well as for consumers. However, it also risks legitimizing and furthering the exploitation of consumers by informal providers.

---

[25] This chapter is published as: Boakye-Ansah, A. S., Schwartz, K. and Zwarteveen, M. (2019). From Rowdy Cartels to Organized Ones? The Transfer of Power in Urban Water Supply in Kenya. The Euroean Journal of Development Research, 1-17.

4.1 INTRODUCTION

The different policies, programmes and projects devised over the past decades to improve access to water and sanitation in African cities have done little to expand services to large segments of the rapidly increasing urban population. The reforms of the 1990s, which were geared towards commercialization, public private partnerships and the opening up of the sector for the introduction of different forms of competition (Jaglin, 2016:186; see also Cosgrove and Rijsberman, 2000; Winpenny, 2003; World Bank, 2003), did not produce the expected extension of the water network, at least not to all areas within cities. Urban water supply in most African cities remains characterized by low coverage and erratic supply, with the wealthier inhabitants enjoying better services than those residing in low-income areas (Water and Sanitation Programme (WSP), 2005). Illustrative of this situation is the Kenyan water sector, where 98 registered water utilities only serve about 55% of the population (WASREB, 2016).

Most water utilities associate low-income areas (LIAs) with widespread illegal water connections, low payment of bills and a high frequency of service disconnection (Heymans et al., 2014: 3). This results in high quantities of so-called non-revenue water (NRW): water that is not formally accounted or paid for. NRW, roughly the difference between water delivered and water billed and paid for, is one of the most important performance indicators for water providers – as it is an indication of the financial sustainability of the service provider (Castro and Morel, 2008; Schwartz et al., 2017). The difficulty in properly accounting for and billing the water delivered to low-income areas is the main reason that most water utilities consider service delivery to these areas a burden (Heymans et al., 2014; Berg and Mugisha, 2010). Subsequently, they prefer to expand services to those areas where tariff collection is easier, and immediate financial returns are more promising (Castro and Morel, 2008: 291).

The resulting gap in formal water supply in African urban low-income areas is often filled by informal modes of water supply (see also Burt and Ray, 2014; Harris and Janssens, 2004; Jaglin, 2008, 2014; Bakker, 2003; Sima and Elimelech, 2011). Where water supply is unreliable or in areas that have rapidly developed and are not (yet) connected to the formal network, informal suppliers thrive (Sima and Elimelech, 2011; see also Angueletou-Marteau, 2008; Batley, 2006; Kariuki and Schwartz, 2005; Misra, 2014). Many of the informal providers engage in water provisioning in LIAs because they consider it an interesting business opportunity. Since their earnings depend on the failure of formal utilities to provide reliable access, they have an incentive to artificially create shortages, by vandalising water pipes or cutting off the water supply to residents (WSTF, 2010). In Kenya, water provided by informal providers is sometimes sold at prices 10 times higher than the official prices to which the formal providers must adhere. Water can be sold for as high as 20 Kenyan shillings (Ksh) (0.19 USD) instead of the recommended 2 Ksh (0.02 USD) for a 20-l gallon, even when the water is of questionable quality.

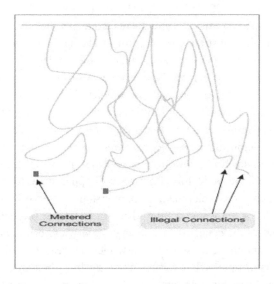

Figure 6: schematic presentation of messy 'spaghetti network' system.
Source: WSP, 2009b.

Informal providers may also obtain their water by illegally connecting to the water network of formal water utilities (WSTF, 2010). They tend not to follow laid down patterns for their distribution networks, resulting in the messy piped network systems sometimes referred to as 'spaghetti networks' (Figure 6). It often happens that these pipes are connected through sewers or laid out on streets and pathways, making them susceptible to frequent breaks and leaks. With the poor sanitary conditions in these areas, the water transported through these pipes also becomes easily contaminated.

In Kenya, the rowdy and exploitative nature of the activities of the informal water operators has earned them the name 'Water Cartels'. Members of these cartels not only include the owners of water points but also the owners of the carts used to transport water to households, cart pushers, and legal or illegal landlords or land- owners. In low-income areas where the formal water utility did manage to develop water provision services, these often ended up being controlled by cartel members. An estimated 55 of such cartels were operating within three LIAs (Obunga, Manyatta and Nyalenda) in Kisumu[26]. As residents have few or no alternatives, other than not having access to water or collecting it from

---

[26] Interview, Staff, KIWASCO, 14 March 2018.

contaminated sources, [27] they remain dependent on the informal water providers. Subsequently, there have been calls to formalize and recognize their activities (De Soto, 2000; Harris and Janssens, 2004; Njiru, 2004; Schaub-Jones, 2008; Solo, 1999; World Bank, 2009). In formulating this call, proponents argue that the recognition will limit the regulatory and financial constraints the informal providers encounter in the course of carrying out their activities (Conan, 2003; Conan and Paniagua, 2003). The ultimate winner of formalization will then be the consumer, who will enjoy better and cheaper service options. In this article, through case studies from low-income areas in three Kenyan cities, where the formal water utilities are partnering with small-scale providers to deliver water to residents, we aim to assess the development of such partnerships in the water sector. We develop our assessment by first presenting the different approaches adopted towards formalization of informal providers in the water sector. We show that, whereas initially the concept of formalization was opposed by formal water utilities, formalization through the development of partnerships to serve LIAs has become an acceptable approach for the water utilities. The article further presents the case studies and ends by discussing the concerns that emanate from these case studies regarding the outcomes of formalization on consumers in low-income areas. Our discussion finds that although the partnership approach to formalization can bring benefits, it also raises strong concerns about the possible exploitation of consumers by formalized small-scale providers operating in partnership with the water utility. Their newly obtained legal status and space does not just make it possible to better control and regulate their behaviours and services, but also allow these providers to wield more influence and power and thus increase opportunities for abuse. This, in turn, can lead to exacerbating the very inequalities that formalization seeks to eliminate.

4.2 APPROACHES TOWARDS FORMALIZATION OF INFORMAL PROVIDERS

Informal providers do not normally conform to the existing rules and regulations that guide formal water delivery in cities, which is why their activities are considered illegal. By such classification, the informal means of water delivery is considered residual or provisional and temporary. With time, they are expected to be gradually replaced by more formal modes of service provisioning (Jaglin, 2016). Unlike their formal counterparts, the illegality of informal providers also means that they are not eligible for financial support from government agencies, donors or banks. By calling them illegal, informal providers were simply not expected to improve and expand their services, as this would make them more permanent.

---

[27] Interview, Staff GIZ, Nairobi, 4 April 2017.

Instead of expecting informal providers to gradually disappear, an approach that gained popularity since the 1990s is to consider their activities as creative and entrepreneurial, and their services as potentially durable. This legalistic approach to formalization argues that, by acknowledging the existence of informal providers and regularising their activities, they become part of the formal market systems (Ahlers et al., 2013; Moretto, 2007; Schaub-Jones, 2008). The approach acknowledges that the informal status of such providers forms a constraint to their longer-term success, while also hindering the improvement of water services provision in areas that formal utilities do not reach. Rather than allowing illegality to thrive (De Soto, 2000), this approach promotes the formalization of informal providers as a way of bringing these small-scale providers into the formal water services system. This approach gained particular prominence in policy literature stemming from the International Development Banks (Solo, 1999; World Bank, 2009).

Formalization through this approach emphasizes that providing legal titles and imposing light-handed regulation to control quality and prices makes it possible for informal providers to have access to adequate financial support (in forms of loans and credits at commercial rates), which allows them to legally invest in their enterprises and access water sources. This in turn is expected to provide them with the incentive needed to reduce the cost of water sold to consumers, as well as to put a halt to their illegal practices in terms of connecting to water sources or networks of formal water utilities. In the longer term, formalization is expected to help unleash a healthy entrepreneurialism in the water sector, by helping markets to emerge in which providers have to compete for customers (Ahlers et al., 2013: 470). This competition is expected to lead to lower prices and also allow consumers to have choices regarding the level of service that they prefer (Johnstone et al., 1999; Njiru, 2004; Solo, 1999; World Bank, 2009).

In sum, the legalistic approach to formalization marks a shift from the government sanctioned monopoly in which formal water utilities are the sole providers of water services (characteristic of the so-called MII) to a market of multiple private suppliers who compete with each other. Lowering the barriers for small-scale providers to formally enter the water services market by creating an enabling environment is the cornerstone of this approach. Formalization thus becomes the route for hitherto 'informal' private providers to become sanctioned as legitimate small-scale entrepreneurs in the sector.

Despite considerable interest and promotion by actors such as the World Bank and internationally operating sector consultants, this approach to formalization never really took off. Formal water sector organizations were reluctant to support it as they saw informal providers as undesirable competitors, with competition moreover happening on unfair terms. Other sector organizations, such as water sector regulators, considered it impossible to regulate such a large number of small-scale providers and highlighted that they had little authority in low-income areas (Ahlers et al., 2013).

In recent years, however, the stance of the producer actors, such as water utilities and regulators, towards the idea of formalizing informal providers has changed. Rather than oppose formalization, they increasingly embrace a new form of formalization. This formalization is not based on granting titles to the informal providers and allowing them to operate in the market, but rather consists of incorporating them into the formal service delivery system through partnership arrangements (Moretto, 2007). These partnership arrangements position the formal water utility as the bulk supplier of water to small-scale providers, who in turn distribute the bulk water obtained from the water utilities to the end-consumers. An important element of these arrangements is that it does not introduce competition between the water utility and small-scale providers. Rather, small-scale providers will have to compete with each other for a share of the low-income markets that the water utility is not able or willing to service (Ahlers et al., 2013; Ranganathan, 2014).

4.3 METHODOLOGY

This study forms part of a larger study currently being carried out in Kenya that seeks to investigate how water utilities address the challenge of ensuring equitable services within polarized cities. Data were collected from 12 LIAs in Nakuru (Kaptembwo, Rhonda and Githima), Kisumu (Nyalenda, Obunga, Otonglo, Nyamasaria, Ogango and Koyango) and Kericho (Matobo, Nyagacho and Mjini). Data were collected through semi-structured interviews with open-ended questions from residents (30), landlords (13), kiosk operators (10) and small/scale providers (8) from these areas. Furthermore, 60 key respondents, including staff of water utilities, staff of local and international non-governmental organisations (NGOs), bilateral and multilateral agencies and government officials operating in the water sector, and staff of governmental bodies were interviewed. In addition, field observations and the review of documents relevant to the study were carried out. Fieldwork was carried out in January to February 2016, January to April 2017 and January to April, 2018. Respondents provided verbal consent to participate in the study, after the purpose and objectives of the study had been explained to them.

4.4 DRIVERS FOR THE PRODUCER PERSPECTIVE ON FORMALIZATION

In 2002, the Kenyan water sector introduced a series of reforms consisting of an upgraded framework for devolution. The reforms addressed two seemingly contradictory objectives. The first objective was to develop the water utilities in Kenya (so-called Water Service Providers, WSPs) into commercially viable organizations. Prior to the 2002 reforms, water supply was under the direct jurisdiction of the provincial authorities and municipalities. Under this arrangement, water supply in cities, especially in low-income areas, was characterized by non-payment of bills and high levels of non-revenue water. For political expediency, authorities did little to rectify this situation. The 2002 Kenyan

Water Act, however, was very much inspired by the concept of commercial public water utilities. WSPs were to operate on the basis of cost-recovery and be financially and managerially ring-fenced in order to avoid political interference in their operations and management. A second objective of the Water Act was to reinforce the Kenyan Constitutional declaration that every Kenyan has the right "to clean and safe water in adequate quantities" (GoK, 2010: 31). The focus on access to water services was further supported by the Kenya Vision 2030 and the commitment to Goal 6 of the Sustainable Development Goals, both of which envisage universal access to water by 2030. The WSPs formed the main tool to achieve this aim.

As these two objectives are not always automatically compatible, water utilities started making a sharp distinction between two customer segments. The need to operate on a commercial basis and recover costs pushes WSPs towards high-income consumers. These consumers are seen as reliable clients who can easily afford tariffs approaching cost-recovery.[28] The need to strive for universal service coverage, on the other hand, forces the utility to direct some of its attention to the large section of the urban population residing in LIAs. These are the very consumers to whom the utilities were reluctant to provide services, because they are considered difficult or recalcitrant.[29] Faced with the dilemma of implementing this 'mixed mandate' (Furlong, 2014) of achieving both commercial as well as social objectives, utilities increasingly saw it as appealing to delegate the responsibility (and the risk) of delivering water to low-income areas to small-scale providers through a partnership approach.

Unlike the legalistic approach where the informal operators are equipped to compete with each other in the water sector, this partnership approach allows the formal utility to maintain a monopoly over water services while delegating part of the less-profitable and more risky activities to small-scale providers. This means that the utility conveniently devolves the actual distribution of water as well as the direct interactions with the consumers of low-income areas to small-scale providers. While thus withdrawing from actual service provision to low-income areas, this approach allows WSPs to take credit for addressing the issue of universal access to services, as they maintain the ability to regulate and control supply through their bulk water provision to the small-scale providers. Moreover, consumers in low-income areas may have specific water demands and willingness to pay for services (Njiru et al., 2001). Partnering with small-scale providers, who are often commended for being flexible, could improve the ability of the utilities to respond to the needs of consumers with appropriate service levels (see Chapter 2). The producer approach to formalization through partnerships thus presents the

---

[28] Interview, Lecturer University of Nairobi, 9 February, 2017.

[29] Interview, lecturer University of Nairobi, 9 February, 2017.

opportunity for utilities to get around the challenges of servicing the low-income areas and balance their financial with their social mandates.

The partnership approach also has an additional benefit in that it 'normalizes' the relationship between the water utility and the Water Cartels, who were previously involved in creating artificial shortages through the vandalism of utility pipes and controlling the local water market. Prior to the formation of these partnerships, WSPs tried different approaches to remove the cartels, for instance by disconnecting illegal networks. This was driven by the fact that the high NRW rates recorded by the water utilities were, at least partially, attributed to the activities of the cartels. The hope was that, after the illegal connections have been taken out of the system, cartel-operated small-sale providers would approach the utilities for them to be connected legally. However, the high connection fee and the desire to continue making money out of the informal provisioning[30] dissuaded both residents and cartels from approaching the utilities for connections. Residents were therefore left without access to clean water – making them vulnerable to water-borne diseases. Formalizing these providers through partnerships instead allows these cartels to be integrated into the formal delivery system. In other words, by partnering with the small-scale providers, the utility seeks to work together with established cartel members who will otherwise 'frustrate' their efforts of supplying water to low-income areas.[31] By embracing formalization through partnerships, the producer actors recognise urban heterogeneity and diversity in demand for water and accept that a singular model to water supply cannot satisfy these demands.

4.5 FORMALIZING INFORMAL PROVIDERS THROUGH PARTNERSHIP

Together with their development partners, WSPs in Kenyan cities introduced the partnership approach to absorb the cartels, involve the community members in the management of their water supply and also further their commercial interests.[32] The hope and expectation is that these partnerships will help control these rowdy cartels, as well as put an end to the unruly behaviour of residents. The latter is expected to happen by allowing them to develop a sense of ownership, so that they assume responsibilities of protecting and maintaining the infrastructure. To summarize these expected outcomes of the approach, an official from a local NGO described it as:

---

[30] Interview, Staff GIZ, Nairobi, 4 April, 2017.

[31] See also Rusca et al., 2015 for an analysis on the institutionalizing corruption in the formal delivery system in Lilongwe, Malawi, through Water User Associations.

[32] Interview, Staff NAWASCO, Nakuru 16 February, 2017.

a bottom–up approach which ensures participation by low-income consumers in water-provisioning issues… it has created room for partnerships between the community and the utility and promotes a sense of ownership among consumers; leading to the reduction of vandalism of infrastructure and the removal of cartels who hitherto were exploiting consumers, thereby making water affordable.[33]

In the sections that follow, we describe three mechanisms that WSPs use to partner with the informal providers in serving the urban poor in three Kenyan cities.

4.5.1 From Cartels to Kiosk Operators

One technology adopted by WSPs to extend supply to the densely populated and often unplanned LIAs is the water kiosk. This technology, which was in use in all three cases, typically consists of a device with one or more taps from which a number of people obtain water. It requires an attendant for its operations. These kiosk attendants provide a good entry-point for the WSPs to form partnerships with existing informal providers. Such partnerships are of two types. Under the first arrangement, the WSP takes care of all the investments required for the construction of the kiosk and then hands it over to an attendant. Typically, attendants are drawn from (1) the informal operators (the cartels) who have prior knowledge of water supply in these areas, and (2) the landlords on whose lands the kiosks are constructed. In the second arrangement, WSPs give permission to individuals or groups to construct kiosks with their own resources.[34] Many private providers whose activities were hitherto classified as informal seized this opportunity to formalize their activities. The cartel members involved in kiosk partnerships are referred to as Kiosk Operators (KOs). They sell water to the consumers and pay the WSPs for the bulk water volumes supplied to the kiosks, as recorded by the water meters at these points.

In line with the partnership approach, the WSPs are responsible for sanctioning prices and overseeing operations at these water points. However, due to a limited ability to oversee and control on the part of the WSPs, the Kiosk Operators are "fully in charge"[35] of the activities at these water points. As most of these kiosks are put up by the operators,[36] they feel entitled to determine the prices of the water they sell at their kiosks. Water is sold between 3 and 20 Ksh for a 20-l container instead of the 1.5–2 Ksh prescribed by the WSPs. In Kisumu, only 12 out of the 60 kiosks visited sold water at the recommended 2

---

[33] Interview, NGO staff, Kisumu, 13 March, 2017.

[34] Here, the utility develops the primary connections and the private operators invest in the secondary pipes and the building of the water point.

[35] Interview, NGO staff Kisumu, 7 March, 2017.

[36] For instance, in Kisumu, only 10 out of the 340 kiosks were constructed by the utility.

Ksh for a 20-l container. At other kiosks, water was sold at 3 or 5 Ksh for the same volume. Prices even go up further to 10 or 20 Ksh when operators deliver water to the doorstep of consumers. Also, in all six kiosks visited in Nakuru and the four visited in Kericho, water was sold between 3 and 20 Ksh, even when the formally regulated price of 1.5 and 2 Ksh for 20 litres was displayed in bold letters on the kiosks (Figure 7).

In making the partnership work, it seems that WSPs pay most attention to the financial aspects of their cooperation. Hence, as long as the kiosk operators pay the required bills, the WSPs do not seem overly concerned by the high water prices they charge. Where these high prices used to be seen as a problem when kiosk operators were still considered informal providers, the partnership with the WSPs has now turned them into a regulatory or enforcement problem. Because of the limited presence of the regulator (The Water Service Regulatory Board-WASREB) in the LIAs, consumers have no other option than to buy water at these high prices. The importance WSPs attach to the payment of bills is evident in the cutting off of supply to kiosks when operators fall behind on payments. This was observed in Kericho where 20 out of the 24 kiosks constructed by the WSP were disconnected due to the non-payment of bills by the kiosk operators. Also, in Kisumu, an average of 15 water kiosks are disconnected at the end of each month for non-payment of bills[37] and the utility has replaced post-paid water meters at water points with pre-paid meters, which requires payment from operators before selling water. Forty-five of such replacements have been carried out at different points in the city.

Riding on the back of these formal partnership arrangements, kiosk operators continue to control the supply of water in LIAs at the expense of the residents. Besides prices, the operators also decide when these kiosks are operational. They open and close access to these kiosks at their own convenience and not when residents need water.

---

[37] These kiosks are reconnected when operators settle their bills.

Figure 7: water kiosk in Nakuru (a) and Kericho (b) showing the price and opening hours.
Source: Author's field note, February, 2016.

4.5.2 Partnerships with Landlords

Another partnership that WSPs adopt as an approach to formalization is with the landlords or landowners in the LIAs, through the construction of yard taps on com- pounds. From the perspective of the WSPs, this is beneficial in two ways. Firstly, it allows the bypassing of the legal challenge related to the lack of land tenure for connections in the LIAs. Even when land tenure is not a requirement for connections, as with the Delegated Management Model (DMM) in Kisumu (which we discuss in detail in the next section), landlords are often reluctant to allow tenants to have their own connections. Secondly, to develop piped networks in the LIAs, WSPs have to consult with the landlords and in some cases pay for laying pipes across their land. Those owners of lands who were already in the 'water business' resisted 'competition' from the formal utilities by refusing to give up their land or charging a very high price for it. The partnership with landowners therefore provides WSPs with an easy way to circumvent this problem. The yard taps, which are constructed either with pre-paid or post-paid water meters, are located on the plots owned by the land-lords, who become responsible for delivering water to 10 or up to 50 households on their compounds.

To the landlords who were already engaged in providing informal water services, the new partnerships afforded the opportunity to formalize what they were already doing on their own compounds.[38] However, once this formality is accorded to them, all decisions on how water is supplied in these compounds also comes to lie with them. This was well

---

[38] Interview, Staff KEWASCO, 27 March, 2017.

articulated by a resident in Rhonda–Nakuru, "everything in this house is decided by the landlord, the only thing you can make decisions about is what happens in your room".[39] The control by the landlords starts with them deciding either to have the yard taps with post-paid meters (where they will be in charge of paying the bills to the WSP) or to have them with pre-paid meters (where residents pay for the water themselves). When they go for the post-paid meters, they decide either to sell water to residents or to add water charges to their rent.

Where water charges are added to the rent, landlords do it at their own discretion. In Nakuru, for instance, landlords who normally do not reside in these com- pounds add water charges of between 100 and 500 Ksh to the rent that residents pay. This means that, irrespective of whether water flows (water supply in these areas is characterized by regular cut-offs, with water being only supplied twice per week for up to 2 or 3 h), residents have to pay. The irregularity and unreliability of the supply forces them to also procure water from other sources – usually other paid water points. In such situations, monies paid as part of the rent are not refunded and residents cannot complain because the only alternative they have is to "find a different accommodation…, which is not a better solution, because the situation is the same everywhere".[40]

In Kisumu and Kericho, landlords prefer to sell water from these yard taps to tenants. They do this to avoid the risk of incurring the extra costs from the over-use or misuse of water by residents when charges are added to rent. Besides avoiding this risk, landlords also consider the re-sale of water as a means to make some extra money. Hence, allowing tenants to pay water charges with their rent or have parallel connections, as one landlord puts it, "will mean losing business…each landlord has to protect his or her territory…houses here are not self-contained where one can have separate things".[41] To increase their earnings from water sales, landlords may extend their water provisioning services to other residents (non-tenants) in their communities. Such yard taps are kept locked when landlord or their agents are not available to sell water to tenants (Figure 8)[42]. Unlike with the kiosks where prices for water are sanctioned by the WSP, landlords in this partnership only have the responsibility of paying monthly water bills, so they control pricing. So important is the re-sale of water to landlords that they delegate this activity to a trusted tenant even when they do not live on the compound. With limited control from

---

[39] Interview, resident Rhonda, Nakuru, 1 March, 2017.

[40] Interview, resident Kaptemboi–Nakuru, 1 March, 2017.

[41] Interview, Landlord, Nyalenda–Kisumu, 17 March, 2017.

[42] It is worth mentioning that yard taps are also locked to prevent unauthorised access from non-tenants.

the WSPs, water, which is delivered to these yard taps at domestic rates, is sold at commercial rates (2 and 10 Ksh) by the landlords in this partnership.

Figure 8: locked yard taps in LIAs.
Source: Author's field note, March, 2017.

4.5.3 From Cartels to Master Operators

WSPs also form partnerships with the cartel members through organisational arrangements. One such arrangement in Kisumu is the DMM. In this model, the utility, Kisumu Water and Sewerage Company (KIWASCO) constructs the primary water pipe connections and installs bulk meters (referred to as master meters) at various points in the LIAs. The final chain of supply, which involves distributing water from the bulk meters to the consumers, is then contracted out to individuals or groups (CBOs), who are called Master Operators (MOs). In this partnership, the responsibilities of the MOs are twofold: to increase and improve connections in the LIAs through individual household connections, yard taps or water kiosks,[43] and bill and collect revenues from connected connections. In total, 39 of such these DMMs are present in 6 out of the 22 LIAs in Kisumu city.

In forming these partnerships, KIWASCO is seeking to improve supplies and their performance in the LIAs by working together with and through the established informal providers. On paper, the MOs consist of groups of individuals who come together to form the CBOs that operate the DMM. In practice, however, MOs usually consist of "one individual; a previous cartel member who registers a CBO with people who may not even exist".[44] Informal providers accept this arrangement, because, contrary to their former operations, registered MOs are accorded the formal status to connect and extend connections to consumers in the LIAs. Also, in this model, land-tenure as a legal

---

[43] Kiosks under this model are owned by individual or groups within the LIAs.

[44] Interview, NGO staff, Kisumu, 7 March, 2017.

requirement for connections is eliminated. This gives residents easy access to water connections.

Besides cartel members gaining legal recognition, this arrangement affords the utility the opportunity to limit NRW rates in the LIAs. This is because MOs pay for every volume recorded by the master meter.[45] They are expected to control or pay for whatever losses that occur within the secondary networks. In the event where the MOs default on payments to the water utility for bulk water, they lose their licence to operate the DMM. This was, for instance, the case of Nyawita (an LIA in Kisumu) where the utility had taken over the operation of all three DMMs due to non-payment of bills by the MOs.[46] The relatively low NRW rate of less than 10% in LIAs served through DMM is an indication that the model has clear benefits for the water utility. To control the leakages, thefts and vandalism, and to ensure that bills to KIWASCO are paid promptly and in full, a number of different measures have been put in place in this model. One of these is the clustering and locking of household-level water meters in chambers (Figure 9). The keys for these chambers are kept by the MOs. According to the official arrangements in the partnership, consumers can ask to check their meters whenever they want to. However, access to these meters is limited. Our interviews revealed that most residents have no idea what the meter records. They are simply expected to pay the bills they are presented with or risk disconnection.[47] When they have genuine concerns about the bills, they have to first pay before they can complain or ask for anomalies to be fixed.

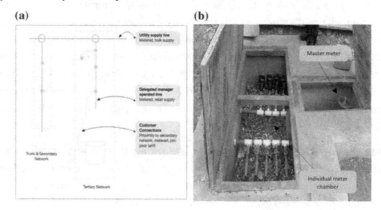

Figure 9: a schematic representation, and b pictorial representation of the DMM. Source: WSP, 2009 and Author's field notes; February, 2016.

---

[45] Interview, Master Operator, Kisumu, 13 March, 2017.

[46] Such DMMs are re-advertised for new MOs to take over after the utility stabilizes the operations.

[47] Interview, resident Nyalenda–Kisumu, 17 March, 2017.

4.5.4 Discussion: Serving the Poor or Satisfying Cartels

Given the difficulties of formal utilities to provide water to LIAs, formalization of the informal sector appears to be a promising solution. Formalization involves a shift from perceiving the informal modes of service delivery as untoward, to pragmatically approaching them as "arrangements that work" (Ahlers et al., 2014: 10). The above discussed partnership mechanisms show that such arrangements indeed seem to present a fit between the seemingly incompatible goals that formal WSPs have to meet: goals of universal access on the one hand and that of financial sustainability on the other. Through the partnerships, the cartels that used to control water distribution and service delivery in the LIAs have been absorbed into and made part of the formally recognized groups responsible for water provision. They become CBOs, or MOs operating the DMMs or KOs. In this way, LIAs have been effectively opened to the WSPs, leading to improvements in some aspects of access. The suspension of land titles as a requirement for extending connections in the DMM and the provision of kiosks and yard taps have made it easier for unserved households to get connected to the network and have regular access to water. The partnerships also lead to improvements in the network structure, and the use of high-quality materials may result in improved water quality. The WSPs also benefit directly from the process. The increase in the number of connections means an increase in their revenues while the reduction of illegal connections and vandalism of the network also translates into a significant lowering of NRW rates. For the WSPs and the regulators promoting such partnerships, they signify a means to extend supplies to 'difficult' areas without compromising commercial viability. For the water cartels, the partnerships offer a way to continue doing their business in a legal manner.

However, the quality of services provided to the LIA consumers raises questions which cannot be ignored. As the literature also shows, formalization in communities with substantial social inequalities may end up aggravating such inequities (Benjaminsen, 2002; Jaglin, 2016; Sjaastad and Cousins, 2009). This is because formalization pro- vides opportunities for the local elites to manipulate water provisioning to their benefit (Sjaastad and Cousins 2009). Indeed, the two lead partners in the newly formed partnerships, the utility and the now formal providers, primarily look out for themselves. The utility has conveniently shifted the risks associated with their services (mainly the risk of not recovering the costs of supplying water) in the LIAs to the MOs, KOs, CBOs and landlords. In the case of the DMM, for instance, the MOs have to protect the infrastructure because any damage or vandalism which results in losses of water means that they have to pay for water they did not sell. The now formal providers also have and feel the responsibility of protecting their investments while maintaining good working relationships with the utility by paying their bills. As evident from the discussion above, they secure their own earnings by shifting all losses to the consumers by charging high prices.

Besides curtailing the activities of the cartels, the partnerships also seek to involve the consumers in their water supply. However, in practice, decisions in the partnerships rest more with the WSPs and their new partners, leaving little or no room for the consumers to contribute. The sense of ownership which these models intended to develop thus remains limited to the members of CBOs, landlords, KOs and MOs. Even between the WSP and the formalized provider, the WSP has the upper hand. For the fear of losing their legality, these providers have to comply with the decisions the WSPs make even when they disagree. The importance the WSPs attach to the payment of bills leaves them with little room to control the activities of the providers in the LIAs. This leaves the formalized provider empowered to control or manipulate the system.

4.6 CONCLUDING REMARKS

With these findings, we question if the real effect of formalization is not the transfer of power from the "rowdy cartels" to organized ones, de facto resulting in a strengthening of their capacity to make profits on the backs of the LIA consumers. Instead of allowing residents to impact decisions, these partnerships seem to be making room for the 'local elites' to wield more influence and wealth, thereby maintaining or even reinforcing the very inequalities the interventions sought to eliminate in the first place. By institutionalizing such an association, with less attention given to the diverse interests of the various actors involved in the (in)formal service sector, inequalities in service provisioning become legitimized. In assuming that informality happens when existing rules and regulations are not binding enough, the intricacies of the informal sector are underrated. By so doing, the sector is viewed as one that can be shaped with ease to fit into the formal arrangements. However, as evident from these case studies, increasing or decreasing formality may not automatically translate into better or worse services for the poor, instead it may mean legalizing informality, leaving the urban poor worse off.

5

THE PRAGMATISM OF ADOPTING PRE-PAID WATER DISPENSERS: INSIGHTS FROM TWO KENYAN WATER UTILITIES

5.1 INTRODUCTION

Pre-paid water dispensers (PPD) represent a technology in which different elements of water supply provision are combined. First, PPDs encompass a meter that records the volume of water dispensed, which is why the technology is also frequently referred to as 'pre-paid meters'. Second, the PPD incorporates a credit-loading system, which reads the amount of credit that a particular consumer has from a token that is used to operate the dispenser. This credit is pre-paid, meaning that the consumer purchases the credit prior to accessing water through the PPD. Finally, the PPD consists of a dispenser that actually provides the water to the consumer. The combination of these three components allows the PPD to dispense a set volume of water to the consumer and to subtract credit from the token in accordance to the volume of water consumed (Heymans et al., 2014).

Since the 1990s, the use of pre-paid dispensers or pre-paid water meters by water utilities serving low-income areas has been the subject of heated debate. On the one hand, PPDs were presented as a tool that would allow utility managers to provide services on a cost-recovery basis, whilst ensuring water services to consumers. This stream of literature argued the potential benefits and advantages of the technology that would accrue to both the water utility and the consumers. The potential benefits for the consumers mainly relate to ease of access and costs. Pre-paid dispensers are viewed as improving access as the pre-paid dispenser functions 24 hours a day, allowing the consumer to fetch water whenever water is needed. The consumer is not dependent on a kiosk operator for accessing water (Heymans et al., 2014). The argument of costs has multiple dimensions. First the absence of an intermediary (such as the kiosk operator) dispensing the water, means that the additional costs of the intermediary are not added to the price charged for water. A second dimension is that application of this technology prevents the consumer from ending up with bills that he/she cannot afford. The consumer knows at all times how much credit is left for purchasing water and can manage water use to accommodate that situation (Heymans et al., 2014). At the same time pre-paid dispensers can benefit the water utility as it allows them to extend water services to consumers without compromising cost-recovery (Chapter 2, Hanjahanja, 2018; Schwartz et al., 2017). The "pre-paid water meter is capable of both measuring the volume of water used, and obligating water users to pay for water" (Berg and Mugisha, 2010:595). By controlling both volumes dispensed and payment, the pre-paid dispenser is presented as "the ultimate cost-recovery tool" (Berg and Mugisha, 2010: 589). In achieving cost-recovery, the technology of the pre-paid dispenser removes utility staff as an intermediary between the dispensing point and the consumer. This is seen as advantageous as it cuts administrative costs and transforms the (political) relationships between the water utility and the consumer (Berg and Mugisha, 2010:595).

A second stream of literature, on the other hand, is critical of pre-paid dispensers. This literature is strongly rooted in a tradition of advocacy against privatization and commodification of water services. For this stream of literature, of which many publications relate to the installation of pre-paid dispensers in South Africa (Deedat, 2002; Loftus, 2006; Von Schnitzler, 2008; Bond and Dugaard, 2010), the pre-paid dispenser represents the ultimate material manifestation of neoliberalization and commercialization. Authors adhering to this critical line of reasoning highlight "the daily indignity and inhumanity" that consumers who have to access water through pre-paid dispensers need to endure (Bond and Dugard, 2010:1). They argue that pre-paid dispensers, which essentially cut off service for consumers if they have insufficient credit, represent a barrier to achieving the human right to water (Mirosa and Harris, 2011), and force consumers to become "calculative agencies, to optimise the household's consumption behaviour and to become economising factors" (von Schnitzler, 2008: 916). Rather than a tool that promotes the agency of the consumer, it thus becomes a tool to constrain their agency. PPDs then become a "water-restricting device" which is used to control the volumes of water that can be accessed by poor households to ensure users pay for the water they consume (Peters and Oldfield, 2005).

Although the two streams are quite different in their focus and argumentation, they are also similar in that they position their discussion of pre-paid dispensers in the larger debate on commercialization. The literature highlighting the benefits of pre-paid dispensers does so by emphasizing how it will improve cost recovery and customer orientation, which are both pillars of commercialization. The more critical literature uses their critique of this technology to underscore a wider criticism of commercialization and the (negative) impact it can have on marginalized consumer groups.

Although we acknowledge the rich contribution of both streams of literature, both deduce a particular set of consequences from the technology. One stream highlights the potential of the technology. The other stream warns for the impact of the technology on marginalized consumers. Both streams do not really engage with why and how water utilities implement the technology in practice. In this article, we seek to contribute to the literature on pre-paid dispensers by examining why utilities opt for PPDs and how they are mobilized in two Kenyan cities (Kisumu and Nakuru). In doing so, we highlight that decisions for opting for a particular water technology, like the pre-paid dispenser, can only be understood in the specific context in which it is implemented. This requires understanding how and why the technology is being introduced and used the way it is.

5.2 THEORETICAL FRAMEWORK: PRAGMATISM

In analysing the adoption of pre-paid dispensers, we argue that decisions made by the water utilities can be viewed as being strongly pragmatic. Although the idea of pragmatism has a long history (see Furlong et al., 2017), the concept has gained increasing attention in the domain of public administration as a reaction to both the traditional model of the bureaucracy as well the New Public Management approach which gained currency in the 1980s and 1990s. In particular, pragmatism criticizes the "one best way orientation" and the "one size fits all views of the world" that these approaches adhere to (Alford and Hughes, 2006:131). Pragmatism "refers simply to an approach in which the organization is open to utilization of any of a variety of means to achieve program purposes, with the choice of these means focused on what is the most appropriate to the circumstances, consistent with important values at stake. In other words, in considering the best management approach to adopt, it all depends on situational factors such as the value being produced, the context, or the nature of the task" (Alford and Hughes, 2006:131).

Citing the work of David Brendel, Shields (2008:205) identifies four dimensions of classical pragmatism, which she refers to as "the four P's framework". In this framework the P's stand for practical, pluralistic, participatory and provisional. These different dimensions of pragmatism are strongly interlinked. Given the nature of the water services sector, being both capital intensive and representing a basic service, we add a fifth dimension related to the positioning of the water provider. The practical dimension of pragmatism refers to the outcomes of decisions made. Classical pragmatism maintains that decisions made ought to provide "practical outcomes for the people in ordinary life" (Brendel, 2006, as cited in Shields, 2008:211). Thus, management decisions should be able to effect transformations, solve problems, clear up difficulties and improve upon existing situation for the people for whom such decisions are made.

Pluralistic refers to the variety of solutions that are available to the decision-maker. Rather than adhering to a 'one best way' approach, the utility manager may try multiple solutions to address a problem. Shields (1996) illustrates this dimension by discussing the 'hotel corridor metaphor'. The idea is that a utility manager wanders through a corridor of a hotel. In the rooms adjacent to the corridor are solutions derived from political science, economics, psychology, sociology and so forth (Shield, 1996). The utility manager will then walk from room to room selecting the solutions that best address the problems that require solving. Given the variety of solutions that the utility manager is likely to select, multiple solutions are expected to co-exist within the organization at a given moment (Alford and Hughes, 2006).

Participatory refers to the idea of involving others in understanding problems and possible solutions. "The open-minded listening, talking, and sharing in a larger context of inquiry sums up the third P of pragmatism" (Shields, 2008:213). Although David Brendel, who developed the framework for psychiatry, originally used participation to refer to the need of the psychiatrist to involve the patient in developing treatment, for our purpose participation refers to water utilities seeking for and investigating possible solutions in their environment, by engaging with other water utilities, key experts, academics and sector professionals. In this sense, participation is also linked by the concept of mimetic isomorphism, in which organizations may adopt solutions that they have seen implemented in other organizations facing similar problems (DiMaggio and Powell, 1983).

The fourth dimension of classical pragmatism concerns the provisional nature of solutions. Solutions are continuously reflected upon and questioned with the purpose of replacing existing solutions with alternative solutions if they are considered more promising and/or more practical. In other words, there is no "one best solution" (Shields, 2008:215), but rather searching for and implementing alternative solutions is a continuous process.

As mentioned earlier, we have added a fifth dimension which we refer to as positioning. Positioning is relevant for water utilities due to the nature of the service provided. In the case of water utilities, the provision of water supply is a basic service with direct public health consequences and thus strongly regulated. Secondly, supplying water is relatively capital-intensive meaning that funding and financing issues play an important role in the sector. Positioning refers to the applicability of the solutions in the context within which they will be implemented by the water utility. Availability of resources and capacity largely dictate what solutions can be realistically implemented. Similarly, government policies and regulations may limit what is doable for utility managers, as they "know full well that there is no escape from the powerful mechanisms of political oversight" (Moore, 1994:296). Shifts in these resources, capacities and oversight mechanisms then also allows for alternative solutions to appear or disappear.

5.3 METHODOLOGY

Fieldwork for this study was conducted in three phases: January to April 2017, January to April 2018 and April to June 2019. In addition to field observations and the review of relevant documents and reports, data for this study was collected through semi-structured interviews with staff of the two water utilities, water point operators and residents from the two cities. Furthermore, key respondents, staff of local and international non-governmental organisations (NGOs), bilateral and multilateral agencies and government officials including staff of the Water Services Regulatory Board (WASREB), the Water Sector Trust Fund (WSTF), and a supplier of pre-paid dispensers (Maji Milele) were interviewed. Before we introduce the two case studies, the article explains the complex institutional environment in which water utilities in Kenya have to operate, forcing them to meet two potentially conflicting policy requirements: that of cost recovery and that of providing water to all. Once this environment has been described, the case studies on the introduction of pre-paid dispensers in Nakuru and Kisumu will be elaborated upon.

5.4 THE MIXED MANDATE OF KENYAN WATER UTILITIES

At the time of Kenya's independence in 1963, water supply issues were treated as part of general government development plan. In this regard, political expediencies and the socio-cultural significance of water were given priority over the need to recover the cost associated with providing the service. With over 80% of the country's population residing in rural areas, the Kenyan Government's first development plan (1964-1970) implemented a policy of free or subsidised water to rural communities, but delegated the supply of water in urban areas to municipal authorities (Mumma, 2007; Onjala, 2002). Governed by the Water Act 1974, water supply was viewed by the Government as the provision of a basic service and not as a source of revenue. Under this arrangement, revenue generated through water rates in urban schemes was projected to constitute only about 40% of the total costs of running the supplies (Onjala, 2002). The remaining 60% had to come from the government and through foreign aid.

Following the economic recession of the 1970s and 1980s, government revenue and the inflow of foreign aid into the country declined. Coupled with an increase in urban population, this made it difficult to continue to justify water supply arrangements which relied on continuous financial support. Kenya's adherence to the Structural Adjustment Programmes (SAPs) of the World Bank and the International Monetary Fund to stabilize its debt situation and as a condition for accessing loans from these bodies further constrained the country's ability to provide free or subsidised water services. This is why the government adopted a new policy position, one that stipulated that "everybody should pay for water services". While continuing to recognise water as a basic need, it was agreed that the operation and maintenance costs of water supply schemes operated by municipalities had to be recovered through user fees (Kenya Development Plan, 1979-83). The new policy statement, and particularly its emphasis on pricing and cost recovery, received a lukewarm welcome from the municipalities. For them, providing water for free or at low cost was important politically and socially. Moreover, their ability to collect fees was limited as only a small segment of those they served paid their bills, the so-called "obedient consumers" (GoK, 1990: 257). A large portion of the urban population, who authorities saw as "recalcitrant consumers", did not pay their bills (ibid). Besides not paying for water, these recalcitrant consumers, many of who resided in the low-income areas of the cities, often also connected to supply lines illegally. Municipalities did therefore not raise enough funds to operate and maintain their own infrastructure, whereas they also did not receive enough financial support to do so. This inevitably resulted in infrastructure breaking down, low coverage and limited access to water for those residing in low-income areas.

This dismal situation triggered the Kenyan Government to change its position on water pricing and cost recovery. In its Fifth Development Plan (1984-1988), it declared that "the provision of, and maintenance of water supply projects will be undertaken primarily through joint efforts between the government and the beneficiaries ... cost recovery will be an essential element in water supply programs" (GoK, 1983: 161). While this declaration signals the introduction of commercial principles into water supply provision in Kenya, it took over a decade for these policy principles to get implemented. It was only in the late 1990s, following the start of the 'privatization decade' in 1993 (Franceys, 2008) that the Kenyan government seriously started commercializing water services provision. It opted for an arrangement in which autonomous water service providers (WSPs) whose activities were independent of the municipal authorities would receive licenses to provide water services. This model was inspired by an earlier pilot by the German Agency for Technical Cooperation, the German Bank for Reconstruction and the Urban Water and Sanitation Management in 1987. As stipulated in the Water Act 2002, the newly created WSPs were to operate "on a commercial basis and in accordance with sound business principles" (Water Act 2002, Section 57(5d)). The emphasis on cost recovery pushed the WSPs towards the more obedient consumers, those who paid their bills. Households in low-income areas were largely left unserved as they presented a risk to the commercial viability of the WSP (Chapter 4). Unserved by the formal WSP, consumers in low-income areas became dependent on the sub-standard services of informal providers (Chapter 4).

Whereas the privatization decade prioritized commercialization and cost recovery, the Human Right to Water (UNGA Resolution 64/292 28 July 2010) and the Sustainable Development Goals (SDGs) revived the policy goals of universal access[48] . Apart from international commitments relating to the Human Right to Water and the SDGs, the goal of universal access is also reiterated to in the Kenyan Vision 2030 and the Kenyan Constitution, which states that "every person in Kenya has the right to clean and safe water in adequate quantities" (Government of Kenya, 2010:31). Yet, achieving the human right to water may contradict achieving goals of commercialization and cost recovery. Whereas commercialization in practice entails a withdrawal of the State from the

---

[48] The first-time goals of universal access were pursued was during the International Drinking Water Decade (1981-1990), which aimed to ensure universal access to water supply by the end of the decade.

provision of services, the human right to water draws the State back in by making it responsible "for both achieving and accounting for progress towards the right to water" (Angel and Loftus, 2017; see also Schwartz and Tutusaus, 2020). The UN Human Rights Council confirms this responsibility of the State by explaining that, "States have the primary responsibility to ensure the full realization of all human rights, and the delegation of the delivery of safe drinking water and/or sanitation services to a third party does not exempt the State from its human right obligations"[49] (Gupta et. al, 2010: 299).

5.5 WATER SERVICE TO LOW-INCOME AREAS IN KISUMU AND NAKURU

The combination of both striving for cost-recovery and being the main instrument for achieving universal service coverage leaves water utilities in Kenya with a difficult 'mixed mandate' (Furlong, 2015: 206): they are expected to achieve both commercial and social objectives. Faced with the challenge of a mixed-mandate, the water service providers in Kisumu and Nakuru have opted for service differentiation to service the population that resides in low-income areas in these cities. Service differentiation concerns a strategy of providing services in which the type of technology to provide services, as well as the associated organisational and financial arrangements to operate that technology, is tailored to different customer segments (Chapter 2). The underlying idea is that consumers in low-income areas 'demand' a different level of service than consumers living in middle and high-income areas (Chapter 2). In the cases of Kisumu and Nakuru, service differentiation also involves an important role for intermediaries in the service provision process. Through arrangements with intermediaries, the water service providers expect that water supply in low-income areas will be characterized by a reduction of illegal connections, an associated reduction of water losses, an improvement in revenue collection, and an improvement in network structure which will result in improved quality of the water supplied (see also Schwartz et al., 2017). Moreover, the pre-payment of water at kiosks means that bills are spread out over time making it easier for consumers to pay their water bills. Furthermore, these bills should be lower as consumption rates from kiosks are lower than for in-house connections.

The Kisumu Water and Sanitation Company (KIWASCO) is responsible for supplying water to the 450,000 people that live in the city of Kisumu. Of this population approximately 60% lives in low-income areas (WASREB, 2019). KIWASCO manages to supply an estimated 76% of the population living in its service area. The main intervention to service six of the 22 low-income areas in Kisumu is the delegated

---

[49] Resolution on Human Rights and Access to Safe Drinking Water and Sanitation (A/HRC/15/L.14), 24 September, 2010.

management model (DMM)[50] . As part of the DMM, KIWASCO constructs the main pipe network from their treatment plant and reservoirs and installs bulk water meters (referred to as master meters) at different points in the low-income areas. For the final stretch of the service delivery process, from the master meter to the consumer, KIWASCO contracts individuals or groups (CBOs). These individuals or CBOs are referred to as Master Operators (MOs) and supply water to consumers through water kiosks, yard taps and individual household connections. Currently, 39 of such DMMs exist in low-income areas in Kisumu. In addition to the DMM, KIWASCO also supplies water to 340 water kiosks, which are located in the other 16 low-income areas. Kiosk operators operate these kiosks.

The Nakuru Water and Sanitation Company (NAWASSCO) is tasked with supplying water to the 510,000 residents living in Nakuru. In Nakuru approximately 40% of the population resides in low-income areas. In servicing low-income areas, NAWASSCO mainly uses water kiosks (40) and approximately 370 yard taps. The yard taps service between 10 and 50 households that rent dwellings from a landlord.

In servicing the low-income areas, KIWASCO and NAWASSCO adhere to three arrangements involving intermediaries. The first arrangement consists of kiosks, which are operated by kiosk operators. The water utilities sell bulk water to the kiosk operator[51] who, in turn, sells the water to consumers. As part of this arrangement, an agreement is made between the utility and the kiosk operator for both the bulk water price charged to the kiosk operator and the price that can be charged to consumers. The second arrangement of co-production concerns landlords selling water from yard taps to consumers who rent dwellings. In this case no formal arrangement exists. Rather the landlord decides how to operate the yard tap and at what price to resell the water from the yard tap to the tenants. Most landlords opt to open the yard tap at specified times so that they are able to control the flow of water through the yard tap. Tenants will either be charged per container filled or through an additional payment added to the rent for the dwelling. The third arrangement concerns the DMM and essentially involves subcontracting the master operators to provide services beyond the master meters. In the DMM, two levels of intermediaries are created. The first level are the master operators who manage the DMM lines. The second level concerns the kiosk operators and landlords who operate the water kiosks and yard tap connections respectively.

---

[50] For a more elaborate discussion of the DMM see Chapter 4 and Sanga and Schwartz, (2009).

[51] These kiosk operators are often drawn from informal operators that were servicing these areas before the water utilities expanded services to these low-income areas.

5.6 THE REASONS FOR INTRODUCING PRE-PAID WATER DISPENSERS IN KISUMU

KIWASCO opted to install 45 PPDs in in 2016. The introduction of the pre-paid dispensers was initiated by the low-income area coordinator of KIWASCO. The motivation for selecting pre-paid dispensers was strongly influenced by experiences presented by other utilities that had piloted pre-paid dispensers. These utilities presented the pre-paid dispensers as a best-practice and highlighted the benefits of this technology, which would allow "public water access […] at the standard tariff" (KIWASCO, 2018). In doing so the technology allows the utility to target both the financial objective of operating on a commercially viable basis and contribute to achieving universal access. Moreover, the technology of the PPDs gave these water utilities a reputation of being "advanced" and "modern"[52]. Of particular importance were the experiences of the National Water and Sewerage Corporation in Uganda and the Nakuru Water and Sanitation Service Company (NAWASSCO). Particularly the "stories of Nakuru fell sweet"[53].

The appeal and promise of the PPDs, as observed during the exposure visits to other utilities, was complemented by the availability of donor funding to procure this technology. The low-income area coordinator of KIWASCO submitted a proposal for 45 dispensers under a project lead by VEI Dutch Water Operators[54] that the water utility participated in. This project, the PEWAK [55] Project, subsidized investments in infrastructure that would lead to improved services to low-income areas.

5.6.1 Removing the Intermediary and Improving Services

In implementing the 45 PPDs, KIWASCO opted to replace meters installed in kiosks with PPDs. The targeting of these kiosk operators for PPDs was part of a strategy to phase out these kiosk operators[56]. By first placing the pre-paid dispensers with kiosk operators,

---

[52]Interview, Staff KIWASCO, 29 April, 2019.

[53] Interview, Staff KIWASCO, 29 April, 2019.

[54] VEI Dutch Water Operators is a Dutch organization, representing five Dutch water utilities that develops international Water operator Partnerships with water utilities in developing countries.

[55] Performance Enhancement of Water utilities in Kenya through benchmarking, collective learning and innovative financing (PEWAK). The aim of the PEWAK project is to improve performance of water utilities in Kenya through benchmarking and peer-to-peer learning, the implementation of NRW reduction plans and through improved service delivery to the urban poor (https://www.vei.nl/projects/performance-enhancement-of-water-utilities-in-kenya).

[56] Interview, Staff KIWASCO, 29 April, 2019.

consumers would get used to the technology of pre-paid dispensers whilst the kiosk would still be operational. In this first phase the kiosk attendants would control the credit-token used to operate the PPD and act as an intermediary between KIWASCO and the end-user. In the second phase a wider distribution of tokens would take place. Each end-user would receive their own token allowing the intermediaries (the kiosk owners) to be phased out[57] .

The reason for wanting to phase out the kiosk owners and attendants was that it would reduce exploitation of consumers by these kiosk operators and landlords (KIWASCO, 2018). This exploitation comes in two main forms. The first form relates to the timeslots during which a consumer can access water. Kiosk operators close their kiosk when they have engagements elsewhere. In other words, operators decided to "open and close water points at their convenience and not when residents needed water"[58] . KIWASCO (2018:4) provides an example of a kiosk operator who "had been out to a funeral for two days and consumers could not access water". The second form concerns the prices charged by the kiosk operators and landlords. The PPD would ensure that not more is charged for safe water than the officially approved tariffs and would lead to a "[r]educed opportunity for fraud by middlemen" (KIWASCO, 2018:2). When water scarcity exists, kiosk operators and landlords have been known to charge more than the official price of Ksh 1.5 per 20 litres. Kiosk operators, in practice sell their water between KES 3 and Ksh 20 per 20 litres (Chapter 4).

By reducing the role of the intermediaries, KIWASCO believed it could provide better services for consumers. "[W]ater is always accessible" as the PPD provides "unrestricted access" (KIWASCO, 2018:2). Moreover, through the PPD, KIWASCO could guarantee that the consumer is able to access at Ksh 1 per 20 litres. This price is even 0.5 Ksh below the official tariff charged for 20 litres at water kiosks.

5.7 THE INTRODUCTION OF PPDS IN NAKURU

The initial decision to opt for the instalment of pre-paid dispensers in Nakuru was influenced by meetings organized by the Kenyan Water Service Providers Association in which pre-paid dispensers were presented as being very successful. In addition to these success stories, possible funding from external donors was also available. The first 15 pre-paid dispensers were installed in compounds housing 6-10 households in 2012 and were financed by the Water Services Trust Fund (WSTF). Each household had their own token to operate the PPD. The experiences of these 15 PPDs were relatively positive as

---

[57] Interview, Staff KIWASCO, 29 April, 2019.

[58] Interview, NGO staff, 15 March, 2018.

consumers were able to access water at all times at a standard tariff set by NAWASSCO[59] . Given this experience, NAWASSCO opted for a second batch of PPDs financed by USAID in 2013. Whereas the first batch of meters served a relatively small community of 6-10 households, the second batch of meters was to serve a larger group of consumers. These PPDs had to serve between 40-50 households. A third batch of pre-paid dispensers was funded by the NGO Water and Sanitation for the Urban Poor (WSUP) and installed in 2014. A fourth batch of PPDs was financed under a project with VEI Dutch Water Operators. An additional series of PPDs were funded by the WTSF in 2016.

The pre-paid dispenser approach taken by NAWASSCO is strongly influenced by the nature of the low-income areas in Nakuru and the issues of water control and payment in these areas. What characterizes the low-income areas in Nakuru is that many people living in these settlements live on gated plots with 20-40 houses that are rented by landlords. Often the landlord is the person who holds the connection with the water utility. The landlord is billed for the water consumed. At the same time, the landlord transfers the bill to individual households by adding an amount to the rent charged for the rent of the house.

The payment of water by tenants leads to a number of problems for these consumers. First of all, the landlord may charge considerably more for water than the amount figuring on the water bill of NAWASSCO. The landlord thus generates a profit on the sale of water to tenants. In Nakuru, water charges added to the rent varied between Ksh 100 and Ksh 500 per month (Chapter 4). A second problem concerned the desire of the landlord to control access to water by consumers. Rather than allow for free access to the water connection, the landlord would lock the tap in order to ensure that tenants would limit water consumption. Despite water prices being added to rents for users of these taps, residents only had access to water for 2 or 3 days in a week – for the rest of the days, taps are locked by landlords[60] .

At the same time, the landlords found the collection of payment from tenants and the need to control access to water a troublesome endeavour. In this context the pre-paid dispenser was thought to be a useful instrument. It would allow the tenant to access water at any time. Also, payment for water would be fixed at Ksh 1.5 per 20 litres, which is below the official tariff of Ksh 2 per 20 litres. Moreover, if a pre-paid dispenser would be placed inside the gated plot, it would safeguard the pre-paid dispenser from vandalism. At the same time the landlord will not be in charge of collection of payment for water and thus will not have to control access to water.

---

[59] Interview, Staff NAWASSCO, 2 May, 2019.

[60] Interview, Staff NAWASSCO, 3 May, 2019.

5.8 DISCUSSION

The framework of the four dimensions of classical pragmatism provides a number of insights into the adoption of PPDs to service low-income areas. Below we explain how the different dimensions apply to the adoption of PPDs.

5.8.1 The Practicality of Solutions

The practicality of the PDDs in the two cases is evident through the desire of the utilities to eliminate the intermediaries (Kiosk Operators in Kisumu and Landlords in Nakuru) who controlled and limited access to water supply in order to make profit and avoid overconsumption. The absence of the intermediaries provides a number of benefits for the consumers. First consumers can have access to water whenever needed and will not have to wait for the landlords to open the taps or kiosk operators to return form their personal errands. Also, the additional cost charged by such intermediaries will be eliminated, consumers will have the opportunity to purchase water at the official price of Ksh 1 as in the case of Kisumu or Ksh 1.5 in Nakuru. The decision for PPDs and the way they were implemented thus aimed at 'practical outcomes' for the consumers living in the LIAs of Kisumu and Nakuru by improving their access to water.

5.8.2 The Existence of Pluralism

The cases of Kisumu and Nakuru reveal an accumulation of strategies and approaches that co-exist simultaneously to service low-income areas. The adoption of a new approach, such as the use of pre-paid dispensers, does not completely do away with 'old' approaches. The 45 pre-paid dispensers in Kisumu, for example, cannot cover all 340 kiosks, meaning that both systems of service provisioning co-exist within the city at the same time. Perhaps more illustrative, however, is that KIWASCO adopted a phased approach to introducing PPDs. In the first phase, KIWASCO did not remove the intermediaries, but rather let the kiosk operators operate the PPD. This meant that in a single kiosk two service delivery arrangements (kiosk and PPDs) essentially co-existed at the same time. In other words, service provision strategies to low-income areas do not involve a single 'best way', but rather a collection of different solutions.

5.8.3 The Importance of Participation

Essential in the decision to adopt PPDs in Kisumu and Nakuru was the role played by other utilities and organizations that provided a positive framing of PPDs as a technology to service low-income areas. In the case of Nakuru, this role was played by the Kenyan Water Service Providers Association. For KIWASCO, visits to NAWASSCO and the National Water and Sewerage Corporation in Uganda, convinced utility managers that PPDs would be a suitable solution for providing services in low-income areas.

Important to note in this context is also the role played by donor organisations in facilitating this learning. For example, the exposure visits for staff of KIWASCO to Nakuru and Kampala, was largely funded by donors. In this sense 'learning' from other actors and water utilities may not always be neutral as the donor needs to agree with what is a desirable exposure visit and what is not. The exposure visits can thus also become a tool for the donor to influence the solutions decided upon by the water utilities[61] . This also hints at a fuzzy boundary between mimetic isomorphism, in which the water utility copies practices of another utility, and coercive isomorphism, in which the water utility is pressured into adopting particular practices.

5.8.4 The Provisional Nature of Solutions

Both KIWASCO and NAWASSCO initially supported the use of intermediaries in providing services. Using intermediaries allowed for a considerable extension of services to low-income areas whilst revenue collection improved and non-revenue water rates dropped to below 10% (Chapter 6; Schwartz et al, 2017;). However, by becoming part of the formal water supply process, the intermediaries, who frequently had origins as informal water operators servicing low-income areas, gained considerable decision-making authority. In the absence of stringent regulation from the water service providers, these intermediaries had wielded considerable power in determining the conditions of providing services in low-income areas. This is perhaps most clearly visible in the prices they charged to consumers and in their control of access to water by opening and closing source points (as described above). This situation was tolerated by the water utilities as it allowed them to meet goals of universal access without compromising commercial principles. Yet, when the opportunity to remove these intermediaries appeared by replacing them with PPDs, both utilities decided to do so. The technology of the pre-paid dispenser provided the promise of removing the intermediary from the provision process, whilst still keeping the risks of servicing low-income areas low. The pre-payment that characterizes these dispensers, means that revenue collection is high and non-revenue water is low, or as an employee of KIWASCO explained "I would just sit and the money comes"[62] . However, also the PPDs may turn out to be provisional solutions. A recent report on diagnosis and repair of PPDs in Nakuru highlights that only 77 out of 242 PPDs

---

[61] Illustrative of using exposure visits to steer solutions adopted by water utilities is the promotion of private sector participation in Uganda by the World Bank. Mbuvi and Schwartz (2010: 380) highlight how "the World Bank facilitated a study tour of key sector stakeholders, including top ministerial functionaries, as part of the Water Sector Reform Study to Ghana, Côte d'Ivoire and Senegal. These countries represented cases where private sector involvement was already in existence or was being pursued".

[62] Interview, Staff KIWASCO, 29 April, 2019.

are actually working (Octrinsic Technologies, 2017 in Kisumu, only 22 out of 45 installed PPDs were functioning in 2019 (Schwartz et al., 2020).

5.8.5 The Positioning of the Water Utilities

The position of the two water utilities is most strongly influenced by the role played by external donors and the Kenyan Government. In both Kisumu and Nakuru, the pre-paid dispensers were either fully procured or heavily subsidized by external donors. Without access to the resources provided by the donors, the water utilities would not have invested in this technology. This was confirmed by one staff member of KIWASCO who posited that: "considering the high cost involved, no utility is interested to invest in pre-paid meters.....however, if a donor comes that wants to support them and install the PPDs, then of course, I will not say no"[63] . Although both NAWASSCO and KIWASCO saw potential and benefits of PPDs, particularly in relation to reducing exploitation by intermediaries, they also indicate that they would not invest in this technology themselves. In other words, they would prefer to adhere to service provision through intermediaries then invest in PPDs.

The institutional environment in which the water utility operates is thus both a source of demands placed upon the water utility as well as a source for solutions. The policy requirements of commercialization and universal access, emanating from national and international policy domains (Tutusaus, 2019) cannot be ignored by the water utilities and largely frame their operations. At the same time the role of donors allows the water utility to access solutions that hold a promise of reducing the possibility that consumers are exploited by intermediaries. The dealings of the Kenyan government and the donors thus determine and shift the boundaries of what is doable by the water utility in the specific context in which it is positioned.

5.9 CONCLUSION

The literature on water services and developing countries has a strong focus on either promoting or critiquing particular technological, managerial or financial approaches to providing water services to low-income areas. In doing so, this literature largely focuses on a 'single approach', either to be implemented or avoided. Relatively little is understood, however, about the way in which utilities make decisions about servicing low-income areas. This article has argued that water utilities adopt a relatively pragmatic approach, accepting solutions they may consider sub-optimal but which, at a particular moment and

---

[63] The unwillingness to invest in pre-paid dispensers was also referred to by other resource persons (Interview, Staff Maji Mele, 6 May, 2019).

within a given context, may be the best they can do. This pragmatic behaviour also leads the operational practices of such water utilities to be relatively messy and unpredictable. The practical limitations and opportunities provided by the institutional environment and the appearance of 'new' alternative solutions that appear as water utilities are exposed to knowledge and experiences of other water utilities, lead to a continuously changing portfolio of strategies and approaches to service low-income areas. Understanding how utility managers pragmatically and continuously manoeuvre and adapt in the context of the complex institutional environment in which they operate in order to provide services is something that remains poorly understood and requires further research.

6

THE APPROPRIATENESS OF 'APPROPRIATE TECHNOLOGIES' IN IMPROVING ACCESS TO UTILITY WATER IN URBAN LOW-INCOME AREAS: EVIDENCE FROM KENYA[64]

In dealing with the challenges associated with water supply in urban low-income areas, water utilities are increasingly adopting service differentiation –making use of so-called appropriate technologies. This article examines the appropriateness of two appropriate technologies; water kiosks and yard taps - in terms of how well they satisfy the water needs of consumers in low-income areas in three Kenyan cities. Based on an analysis of the reliability, affordability and accessibility of water supplied through these two technologies, we conclude that while service differentiation has led to improvements in access to piped connections in low-income areas, it also risks reinforcing the inequalities in access to water.

---

[64] This chapter is published as: Boakye-Ansah, A. S., Schwartz, K. and Zwarteveen, M. The Appropriateness of "Appropriate Technologies" in Improving Access to Utility Water in Urban Low-Income Areas: Evidence from Kenya. Submitted to (under review) International Journal of Water Resources Development.

6.1 INTRODUCTION

In spite of the progress made towards improving access to drinking water over the past decades, significant challenges remain. Globally, about 700 million people still lack access to improved or clean water sources. A large proportion of this population is located in the sub-Saharan region of Africa (UNICEF and WHO, 2015). In this region, the effect of poor access is especially severe in low-income or informal settlements within cities. In Kenya, for instance, while 84% of households in high and middle income settlements have access to water through piped sources, this is true for only about 34% of households in low-income areas within these cities (World Bank, 2015). Residents in these areas instead rely on informal water providers and other unimproved sources for meeting their water needs.

The lack of service provision in low-income areas is the result of a number of factors. Rapid population growth coupled with the unregulated and haphazard manner in which houses are built make it challenging for water utilities to develop water infrastructure in these areas. Water utilities are also often legally unable to extend the network into places where the inhabitants lack land-tenure. Even if they would be allowed to extend services here, water utilities are hesitant or reluctant to do so, because it is fraught with challenges and risks. These include the lower ability of consumers to pay for water services (Cross and Morel; 2005); low water consumption levels (Schwartz et al., 2017); high rates of vandalism; a high number of illegal connections; and high non-revenue water (NRW) rates (Bakker et al., 2008; WSP, 2009b). Together, these challenges and risks make it difficult for utilities to provide services to low-income areas in cost-effective ways. This is problematic, as utilities are not just under pressure from donors and governments to ensure universal coverage, but they are also expected to provide services in (quasi-)commercial ways, on the basis of cost recovery (Chapter 2; Schwartz et al., 2017).

An increasingly popular strategy of water utilities to combine the goals of cost-recovery with that of universal coverage is to resort to what has come to be known as service differentiation and the use of so-called appropriate technologies, with services provided and technologies used in low-income areas being different from conventional ones. Hence, in low-income areas appropriate technologies are to be used instead of the on-premise connections that characterize the middle and high-end areas of the cities. In this paper, we examine two of the appropriate technologies (water kiosks and yard taps) that figure prominently in service differentiation strategies. We look at the impact of the use of these technologies on access to water and assess their appropriateness in satisfying the water needs of consumers. We begin our analysis with a more general reflection about service differentiation and the use of the associated appropriate technologies for water supply in urban low-income areas.

6.2 SERVICE DIFFERENTIATION IN URBAN WATER SUPPLY

In recent decades, policy actors in the international drinking water and sanitation sector have promoted service differentiation as an approach to address the lack of water services provision in low-income areas (Mara and Alabaster, 2008). Central to the idea of service differentiation is that it is difficult (for a range of reasons) for water utilities to provide the same services to all consumers. As such, service differentiation entails an acceptance of the existence of social difference. Consumers are not all the same: they can be divided in distinct categories on the basis of geographical location and income, with services being tailored specifically to each category. The one distinction that matters most in service differentiation is the one between willingness or ability to pay, or that between low-income users or areas and others. Service differentiation profiles residents in low-income areas as more unreliable when it comes to paying the connection fees and monthly water bills, because many of them are employed in the informal sector, and have low and fluctuating income levels (Berg and Mugisha, 2010). To be able to provide water to these areas and users without running the risk of incurring losses, water utilities tailor water provisioning to the specific needs and demands of low-income consumers through service differentiation (Jaglin, 2008). In practice, this means providing a lower (or more 'appropriate') level of water service than what is provided to everyone else through conventional in-house connections.

To do this, utilities resort to different technologies, those that are more 'appropriate'[65], to deliver water to low-income areas. These include water kiosks, yard taps or pre-paid dispensers. What makes these technologies more appropriate for the water utilities is the particular organizational arrangements and financial regimes associated with them. These arrangements simultaneously reduce consumption levels and facilitate the collection of fees. Unlike with the conventional in-house technologies where management lies with the water utilities, with appropriate technologies, the management or organisation of service delivery is delegated to intermediaries. Depending on the type of technologies such intermediaries can either be human (small-scale providers) as in the case of the water kiosk and the yard tap or technological intermediaries as in the case of such as pre-paid water dispensers (Chapter 5). With the human intermediaries, a local person is appointed by the utility to operate a lockable standpipe and is responsible for charging residents for each container of water they collect from these points (Loftus, 2006). Such intermediaries come in the form of single individuals as in the in the case of the bailiff system in Durban South Africa or the Kiosks Operators system in Kenya or in the form of groups as in the case of Lilongwe (Malawi) where the technologies are entrusted to water user

---

[65] The term 'appropriate' is used frequently in this article. As it would be rather irritating to write the word in quotation marks every time, we have opted to use the term without quotation marks.

associations (WUAs) who then appoint operators to sell water at these water points (Chapter 4; Loftus, 2006; Rusca et al., 2017). With most residents in the low-income areas lacking land tenure, the legal requirement for acquiring service contract for conventional in-house connections and with the high cost burden of such connections for the residents, such delegation allows the utilities to provide water to these areas whilst ensuring that water is paid for. The financial regime of appropriate technologies often takes the form of pre-payment: where money is paid directly to the human intermediaries or consumers pay through their tokens when accessing water from a PPD. Prepayment is argued to promote financial 'self-sufficiency' and allows households to pay for water when needed and for the specific amounts used (Heymans et al., 2014). In promoting service differentiation "differentiating between different categories of households and treating citizens differently according to different circumstances," is accepted as the level of service enjoyed is linked to an associated level of payment. The lower the level of service received by a consumer, the lower the price charged to that consumer (Van Ryneveld et al., 2003:43). In other words, the idea is that the technology, the organizational arrangement and financial regime through which water services are provided better match the demands of the consumer groups (Mara and Alabaster, 2008; Njiru et al., 2001). High-income groups, who demand a relatively high level of service also pay more for this level of service. Low-income households that are supplied through appropriate technologies, would be expected to pay less for the lower level of service they receive.

Although the approach is widely embraced by utilities, governments and donor organisations, there are many critical questions being raised about how equitable and 'demand-responsive' service differentiation really is (Chapter 4; Gulyani et al., 2005; Lauria et al., 2005; Schwartz et al., 2017). First studies in Kenya and other places highlight the low consumption levels associated with the technologies, sometimes as low as 3 litres per capita (Adank and Tuffour, 2013; Chapter 4; Hadzovic, 2015; Rusca et al., 2017; Schwartz et al., 2017). These studies further show that prices of water (per cubic meter) supplied through appropriate technologies are often higher. If indeed the use of appropriate technologies does not lead to reliable and affordable services for low-income consumers, the question is warranted; how appropriate are these technologies and, for whom they are appropriate? Indeed, the critiques suggest that there may be a mismatch between the arguments that promote service differentiation and their effects and impacts of their use in low-income areas.

The remainder of this article engages with the emerging discussion about service differentiation and the associated appropriate technologies. We address these questions by analysing service delivery through water kiosks and yard taps in low-income areas in three Kenyan cities (Kisumu, Nakuru and Kericho). We assess the appropriateness of the water service delivery systems for low-income households by examining three important dimensions of water access: affordability, reliability and accessibility of water services. Based on the results of our analysis of the two technologies, we conclude that while service differentiation may lead to improvements in access to water in low-income areas, it also risks broadening existing inequalities in access to water. This is because the financial and organisational arrangements associated with the implementation of appropriate technologies tend to make access to water more difficult and costlier for consumers in low-income areas. The net result is that consumers in low-income areas pay more for poorer services, or are prompted to resort to unimproved sources which are usually unfit for consumption. These findings suggest that the financial and organisational arrangements that accompany appropriate technologies are above all appropriate for water utilities and the intermediaries rather than being appropriate for the consumers in low-income areas to whom they are targeted. The use of appropriate technologies allows the water utility to delegate costs and risks associated with water services provision to low-income areas to intermediaries, and these intermediaries transfer such costs to consumers. Even when the utility supplies water at subsidised rates to the intermediaries, their profit-seeking orientation pushes them to charge higher prices to the consumers. The subsidies granted by the utilities, then, mostly benefits the intermediaries

6.3 METHODOLOGY

6.3.1 Study Area: Institutional Set-Up for Urban Water Supply in Kenya

The water services sector in Kenya involves a number of different organizations that either provide water services, facilitate provision or regulate water service delivery to low-income areas. Established by the Water Act of 2016[66], the Water Sector Trust Fund (WSTF) has the mandate "to assist in financing the development of and management of water services in the marginalised and underserved areas"[67] . The Water and Sanitation Regulatory Board (WASREB) regulates the Kenyan water services sector. According to the 2016 Water Act WASREB's "principal object is to protect interests and rights of consumers in the provision of water services" (Water Act, 2016). In order to pursue this objective WASREB "sets, monitors and reviews rules and regulations to ensure water services provision is affordable, efficient, effective, and equitable"[68] . WASREB issues licenses to Water Service Boards (WSB), which are responsible for managing water infrastructure assets. The WSBs are also obligated to ensure that all consumers have access to sustainable and affordable water services (Schwartz et al., 2017). As asset

---

[66] The Water Sector Trust Fund, previously existed as the Water Services Trust Fund under the Water Act of 2002.

[67] https://waterfund.go.ke/. Accessed 7 January, 2020.

[68] https://wasreb.go.ke/about-wasreb/. Accessed 7 January, 2020.

holders, the WSB transfer responsibilities for actual water provision to water utilities, called Water Service Providers (WSPs), through Service Provision Agreements. As such, in the Kenyan water services sector, responsibilities of regulation, asset holding and provision have been delegated to different organizations.

6.3.2 The Case Studies

The study was conducted in sixteen LIAs located in three cities in Kenya – Kisumu, Nakuru and Kericho. These sixteen LIAs were typical of LIAs in Kenya where supply of piped water is limited and residents have to rely on alternative (unimproved) water sources, including (shallow)ground water, rain water and surface water to argument their water needs. These LIAs were selected because they have seen the introduction of appropriate technologies in the recent past. Hence, residents of these areas were supplied water through water kiosks and yard taps.

Kisumu, which is located on the shores of Lake Victoria, is Kenya's third largest city. About 60% of its population of approximately 500,000 live in twenty-two LIAs. Two LIAs, Nyalenda B that includes the clusters Western B, Kilo, Dunga, Got Owak, and Nanga and Nyamasaria were selected for the study. Here, Kisumu Water and Sewerage Company (KIWASCO) is the WSP mandated to provide water for the city's inhabitants (Owuor and Foeken, 2012). KIWASCO supplies water to the LIAs through a so-called Delegated Management Model (DMM). Developed in 2004, the DMM entails that KIWASCO delegates the last chain of supply in the LIAs to small-scale operators; so-called Master Operators (MOs). The MOs have the responsibility of connecting users either through water kiosks or through the yard taps[69]. They pay KIWASCO for the bulk water it supplies (Chapter 4; Nzengya, 2017) and recover this money through collecting fees from the customers that they serve.

Nakuru, the fourth largest city in Kenya with a population of approximately 500,000 (NAWASSCO, 2013) is one of the fastest growing cities in the east African sub-region. Approximately 40% of the city's population live in 40 LIAs (Chakava and Franceys, 2016). In Nakuru, data for this study was collected from ten low-income areas (Kaptemboa, London, Kivumbini, Nakuru Railway, Kambi Somali, Manyani, Pangani, Barut, Wangamata and Gilani). In 2003, the Nakuru Water and Sanitation Company (NAWASSCO) was given the mandate to operate water and sanitation services in the city (NAWASSCO, 2013). Unlike in Kisumu, where the WSP delegates water supply in the

---

[69] This model was first piloted in 2004, in Nyalenda (the biggest LIA/informal settlement in Kisumu) with four MO lines. Currently 39 DMMs exit in 6 out of the 22 LIAs.

LIAs to the MOs, NAWASSCO itself directly operates in the LIAs. The company has a dedicated pro-poor unit that oversees the company's operations in the LIAs[70].

Kericho with a population of approximately 180,000 is the largest town in the Kericho County. About 34% of this population live in the ten (10) LIAs in the town (KEWSACO, 2018). Water supply is carried out by Kericho Water and Sanitation Company Limited (KEWASCO service charter, 2007), a company which started its operations in 2002 after its incorporation in 1997 (KEWASCO service charter, 2007). Four LIAs (Brooke, Kapsoit, Kapsuser and Nyagacho) were selected for study in Kericho. Just as in Nakuru, operations in the LIAs are carried out directly by the WSP, with the dedicated pro-poor unit overseeing the WSP's activities in these areas.

6.3.3 Study Methods

The fieldwork for the study was carried out in two phases by the main author; from January to April 2017 and from January to April 2018. The main method used was a survey based on a structured interview of randomly selected 515 households who were users of kiosks and yard taps and 37 kiosks operators (Table 7). A household in this context was defined as a group of people living together and sharing domestic resources. The household size in this study ranged from two to twenty, with an average of $5(SD=1.04)$.

The survey sought to gather information on the household income and percentage of that income spent on water, water use practices, the type of appropriate technologies used for collecting water, and the number of days and hours in a day water is supplied by the WSPs. For household income, the survey relied on self-reported income from respondents. Where households could not report exact income figures, they were asked to indicate which of the following five daily income brackets they were in: less than or equal to Ksh 100, Ksh101–500, Ksh501–1,000; Ksh1,001–1500; Ksh1501–2000. This income bracket was based on the minimum income/wage set by the government of Kenya for urban dwellers. In order to reduce survey bias on income earned and amount of money spent on water, survey questions were posed three times; on a three-month interval. Average figures from these responses were then recorded.

Following the survey, the main author conducted semi-structured interviews with Master Operators (n=7) and residents and also carried out direct field observations. Documented evidence such as water bills, maintenance or water supply reports were collected to support verbal responses, when available. Questions were asked in Swahili, the predominant local language across Kenya and in Luo and Kipsigis, the tribal languages in Kisumu and Kericho respectively. Participation in the study was voluntarily, and

---

[70] It is worthy of mentioning here that pro-poor units/departments exist in all three WSPs.

respondents could drop out at any point. Respondents gave their formal oral consent after they had been duly informed about aims and purpose of the study.

Table 7: Distribution of respondents per selected LIA

Study area		Consumers/ Residents	Kiosk Operators	Master Operators	TOTAL
Kisumu	Nyalenda	104	20	5	129
	Nyamasaria	91		2	93
Nakuru	Kaptemboa	40	7		47
	London	20			20
	Kivumbini	20			20
	Nakuru railway	20			20
	Kambi Somali	20			20
	Manyani	20			20
	Pangani	20			20
	Barut	20	2		22
	Wangamata	20			20
	Gilani	20	2		22
Kericho	Brooke	40	2		42
	Kapsoit	20	2		22
	Kapsuser	20			20
	Nyagacho	20	2		22
TOTAL		**515**	**37**	**7**	**559**

Source: Author.

6.4 RESULTS

6.4.1 Water Technologies

The water kiosk

The water kiosk typically consists of a structure located outside of the dwelling of users, equipped with one or more taps from which a number of people obtain water. Two types of water kiosks were present in the study areas. The first type consisted of booths that can be used by the operators to pursue other commercial activities in addition to selling water. These booths are put up by the water utilities, through financing from different donor organisations (Figure 10). A second type of kiosk is one that consists of a single pipe constructed by individuals or groups, with authorisation from the water utility (Figure 11). For both types of kiosks, users are required to pay before collecting water, either to the attendants at these points (referred to as Kiosk Operators) or through tokens in case the point has a pre-paid meter installed (as in Kisumu and Nakuru). Of the 195 respondents in Kisumu, 4% (n=8) made use of water kiosks, whereas 19% (n=41) and 33% (n=33) of the 220 respondents in Nakuru and the 100 respondents Kericho respectively collected water from this source. The Water Services Regulatory Board (WASREB) recommends that each kiosk serves a maximum of 250 people per day[71]. This corresponds more or less to what is happening in Kisumu, with each kiosk serving up to 200 people per day. Yet, the number of people served per kiosk is higher in Nakuru and Kericho. In Kericho, kiosks serve up to 300 people, whereas in Nakuru one kiosk can even serve up to 500 persons.

Figure 10: water kiosk constructed by the water utility.
Source: Author's field notes: 22 March, 2018.

Figure 11: water kiosks constructed by groups (on the left) and by individuals (right).
Source: Author's field notes; 25 February, 2018.

---

[71] This is based on the assumption that, each household is made up of five members, and that each kiosk serves up to 50 households.

Yard tap

Yard taps are piped connections to a tap placed in the yard or plot outside of the house (Figure 12). They either serve as the water source for single families or are shared among a number (up to 50) of households. Most of the respondents in this study collected water from this source. In the case study areas respectively 96% of respondents (Kisumu), 81% of respondents (Nakuru) and 67% of respondents (Kericho) accessed water through yard taps.

The payment of the water bills for these connections happens in the same way as for the conventional connections in other parts of the city. Landlords of the compounds on which a tap is placed have the responsibility of paying the bill to the water utility at the end of each month. They fulfil this responsibility in two ways. The first involves landlords of plots or compounds adding the cost of water to the rent of tenants. Landlords add water charges of between 100 and 500 Kenyan shillings (Ksh)[72] to the monthly rent that residents pay. This was the case for all the yard taps we surveyed in Nakuru. The second involved the landlords selling water from these taps to members of households within the compounds, as well as to neighbouring households. This was the case for the yard taps surveyed in Kericho and Kisumu

Figure 12: example of yard taps.
Source: Author's field notes; 25 February, 2018.

6.4.2 Affordability of Water Supplied

To understand water affordability, we assessed the cost of water as a percentage of the incomes of the households in the study area. For water to be affordable, the United Nations recommends that the cost should not exceed 5% of a household's income (United

---

[72] 1 Kenyan shilling was equivalent to 0.0098 USD.

Nations, 2010). The affordability levels presented in Figure 13 show that some of the households we surveyed spent a greater proportion of their income on water.

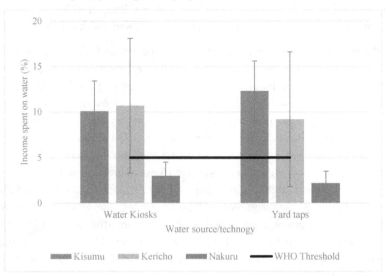

Figure 13: affordability of water across the different sources.

Source: Author.

The survey further revealed that, households paying monthly water bills as part of their rent as is the case for yard taps in Nakuru paid less than households paying for water on case-to-case basis either from water kiosks or from yard taps in Kisumu and Kericho. In such households in Nakuru, water cost translated to 0.06 Ksh per litre. This cost does not only correspond with the water charges in the middle and high income areas of the city but it is also in line with the official domestic tariffs charged by WSPs for the first 6000 litres of water used by consumers. However, cost per litre for water collected from kiosks translated to 0.2 Ksh in Kisumu, 0.1 Ksh in Nakuru and 0.3 Ksh in Kericho. Similarly, the cost per litre for water collected from yard taps in Kisumu and Kericho translated into 0.2 Ksh.

The main reason why consumers in Kisumu and Kericho spend so much of their income on water is that the price charged by the intermediary (kiosk operator or landlord) is relatively high. The WSPs supply water to the kiosks at a bulk rate of about 35 Ksh per cubic metre[73] . The WSPs then expect Kiosks Operators to charge customers between 1.5-2 Kenyan shillings (Ksh) for a 20-litre container. However, in practice the Kiosk Operators sell water at prices fluctuating between 2 and 5 Ksh per 20-litre container, with

---

[73] This translates to 0.7 Ksh per 20 litres.

prices sometimes even going up further to 10 or 20 Ksh when operators deliver water to the doorstep of consumers. Similarly, water from yard taps in these low-income areas is billed by the WSPs at the domestic tariff of 0.06 per litre, or 1.2 Ksh per 20 litres. However, it is sold by the landlords for prices between 2 and 5 Ksh per 20-litre container.

Besides the price of water at the different sources, the low and often fluctuating income levels of residents also compromise the affordability of water in these areas[74]. Income levels ranged from a minimum of 120 Ksh to a maximum of 1440 Ksh, 30 Ksh to 1000 Ksh and 100 Ksh to 700 Ksh per day for households in Nakuru, Kericho and Kisumu respectively. In Kenya, the minimum income/wage set by the government for urban dwellers is a 269.40 Ksh per day for unskilled labour and 1404.30 Ksh per day for skilled labour. While the income levels for most of the respondents in Nakuru (97% of the 220) conformed to the government's stated minimum (for unskilled labour), income levels for only about a half of the households in Kisumu (55% of the 195) and Kericho (54% of the 100) were in line with this requirement. The higher income levels coupled with the relatively lower water prices ensured that water was more affordable for households in Nakuru in comparison to those in Kisumu and Kericho. The actual percentages of households that complied with the UN's affordability threshold for each of the cases is presented in Figure 14.

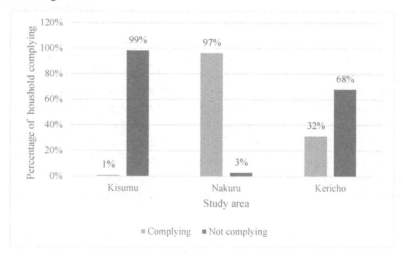

Figure 14: percentage of households complying with the affordability threshold.
Source: Author.

---

[74] Often, incomes for most of the households (about 72% in this study) was earned on daily basis and so fluctuated depending on the availability of work.

6.4.3 Reliability of Water Sources

The reliability of the water sources was measured by the number of days and hours that water was available to households in the LIAs studied (Figure 15). In relation to piped water sources, the benchmark by the UN is a minimum 12 hours of supply per day (WHO and UNICEF, 2017)

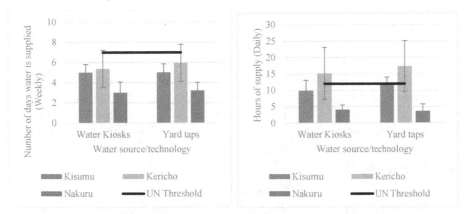

Figure 15: reliability of supply at the water sources; left: number of supply days and right: hours of supply per day.
Source: Author.

With the exception of Kericho, where officials of the WSP (KEWASCO) reported rationing both KIWASCO and NAWASSCO reported a 24 hour, 7 days a week supply. Yet, the results presented above show that there are frequent interruptions in the supply of water in all cases – with Nakuru recording the lowest reliability. According to respondents, these interruptions are not only because the WSPs do not supply bulk water to the kiosks and yard taps, but also because of the unavailability and 'tap-opening restrictions' of kiosk attendants, MOs and landlords.

Unlike the conventional in-house connection where the responsibility for the reliability of supply rests with the WSP, the supply of water at the water kiosks and yard taps depends largely on the availability of the attendants and the landlords at these water points. In Nakuru for instance, to ensure that residents do not misuse or over-use water, landlords lock the taps and open them for an average of three days per week for up to four hours on each of these days. To ensure continuity of access, residents in these compounds collect and store water for later use. When they run out of water before the next "tap opening day"[75] , they have to get water from other sources. In addition, kiosks are not always attended to; attendants are frequently engaged in other commercial activities and are not

---

[75]Interview, Resident Nakuru, 12 April, 2018.

present to serve consumers at all times. Other issues such as family emergencies also compromise the availability of the attendants. Illustrative was a kiosk in Nyalenda in Kisumu, which was closed for a week because the attendant had to attend to her sick mother in another part of the country. For fear of mismanagement of monies collected at the kiosks, the attendants prefer not to delegate the selling of water to someone else in their absence.

Besides the availability of the attendants, leakages and other damages to water pipes in these LIAs were also reported by respondents as a factor contributing to the unreliability of supply at the water sources. Water pipes in these areas are often exposed on roads and walkways. With the pipes exposed, they become susceptible to damage through accidents or vandalism by residents – resulting in the limiting or the cutting off supply. Such a situation was observed at a kiosk in Nyalenda, where water supply had been cut off for days due to the destruction of the connecting pipes because of a road construction works.

6.4.4 Accessibility

We measured accessibility as the total time spent in collecting water from the sources. This consisted of the time spent walking to and from the source and time spent waiting to collect water at the source. For improved access, the UN recommends a maximum collection time of 30 minutes. To establish a total collection time of 30 minutes, the WASREB, in addition to limiting the number of people a source should serve also recommends that water sources should be located not more than 200 meters (0.2 km) from dwellings. Results from our survey (Figure 16) show that, with the exception of Nakuru where the time household members needed to collect water from yard taps is within the recommended limit, collection time for Kisumu and Kericho exceeded this limit. Households however, spent less time collecting water from yard taps than from water kiosks. The actual percentages of respondents complying with the distance and total time thresholds are presented in Figure 17.

Our survey found that waiting time formed the largest proportion of the total time spent on collecting water. While households spent an average 5.51 minutes walking over the various distances to their water sources, they spent an average 25.79 minutes waiting to collect water from kiosks or yard taps. This longer waiting time is a direct result of the large number of people making use of these facilities, often resulting in long queues. Low water pressures also increased the waiting time.

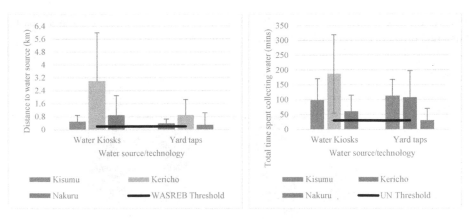

Figure 16: accessibility to water at the sources; left: distance to water source and right: time spent collecting water.

Source: Author.

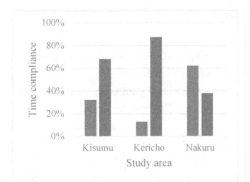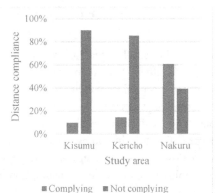

Figure 17: percentage respondents complying with accessibility threshold: left: time compliance and right: distance compliance.

Source: Author.

6.4.5 Implications on Water Consumption

In this section, we comparatively analyse the water consumption levels among households in the study areas. Further, we show how affordability, reliability and accessibility of water from the different sources impact water consumption levels among households.

In non-emergency situations, the WHO recommends a daily per capita consumption level of 50 litres for drinking, cooking and other hygiene purposes (WHO, 2003). With the exception of households in Nakuru, the amounts of water consumed by households in Kisumu and Kericho were below this level (Figure 18).

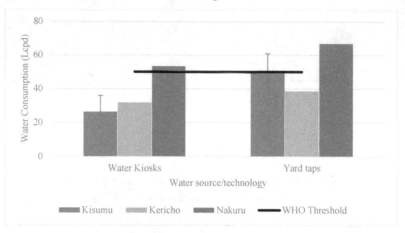

Figure 18: average water use among respondents.
Source: Author.

For many households, the volumes of water collected from kiosks and yard taps were not enough for their daily domestic activities. Yet, because they are already spending a significant proportion of their income on water from kiosks and yard taps, many indicated preferring to resort to other (unimproved) sources to fill the gap. Members of these household used utility water mainly for cooking and drinking, and relied on other, cheaper sources of water – including surface water (rivers, stream and springs), rainwater as well as (shallow) wells for other domestic activities such as laundry and bathing. Even those households with the lowest consumption levels spent a considerable percentage of their income on their water needs (Figure 19).

Besides affordability, the time needed to collect water from kiosks and yard taps influenced how much utility water people consumed (Figure 20). Especially the long waiting time makes those household members responsible for collecting water limit their visits to water kiosks or the yard tap. Instead of waiting in long queues to collect water, they prefer to collect water from other more easily accessible sources. Collecting water from these sources, according to a resident in Kericho, "saves them time to engage in other profitable ventures"[76] . For these residents who are often employed in the informal sector, the time saved goes a long way in improving their income levels.

---

[76] Interview, Resident Kericho, 18 March, 2018.

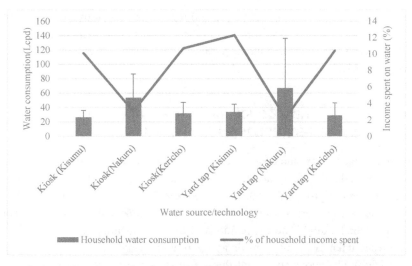

Figure 19: relationship between affordability and water consumption.
Source: Author.

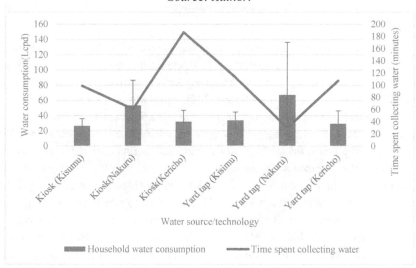

Figure 20: relationship between times spent at water sources and consumption levels.
Source: Author.

The limited availability and the supply intermittences also reduced consumption levels among households. In times of water shortages or in the event of the closure of the water points by the attendants or landlords, households have no option than to resort to alternative forms of water provision. The cost of storage containers and the space needed to store these containers also limits the volumes of water that residents have available for later use. According to residents, the price of these containers – which ranges from 100 ksh for a 20 litre jerry can to 1100 ksh for a 200 litre container – adds an extra cost to the already high price of water. As dwellings in these settlements often consist of a single housing unit used by at least five members, finding extra space to store water containers also become difficult.

6.5 DISCUSSION AND CONCLUDING REMARKS: FOR WHOM ARE APPROPRIATE TECHNOLOGIES APPROPRIATE?

Development practitioners and governments promoting service differentiation and the use of appropriate technologies argue that, in the context of polarised cities in developing countries, this model provides a more flexible room for water utilities to provide water services for low-income areas while maintaining their commercial/cost recovery mandate (Castro and Morel, 2008). The expectations of the implementing utilities are that such technologies will remove limitations on access to utility water supply in the low-income areas. Survey results above show that the use of such appropriate technologies indeed seems to contribute to the expansion of piped network in the low-income areas. At the same time, however, the three cases studied, highlight the limitations of appropriate technologies.

A key characteristic of the appropriate technologies studied in the three Kenyan cities concern the organizational arrangement and associated financial regime through which the technologies were being used. In all cases service delivery responsibilities were delegated to intermediaries (kiosk operators, MOs, landlords). This had the advantage for the utility to transfer the financial and operational risks associated with service delivery in low-income areas. By focusing on bulk water provision to a limited number of kiosks and yard taps, the water utility basically minimizes its financial risks. However, the involvement of the intermediaries comes at a considerable cost for the consumers in the low-income areas. In two cases the intermediaries significantly increased the prices charged to consumers for drinking water. Prices charged to consumers were between 3 and 7 times higher than the tariff charged by the WSP to the intermediary. In some cases, prices were even 28 times higher. With limited household income, the people dependent on these technologies tend to focus on meeting their more pressing needs such as rent and food and reduce the volume of potable water they collect from the paid sources to a minimum. Free, unimproved, alternative sources are used in order to avoid having to spend a larger part of the income on water.

Also, in terms of accessibility, the role of the intermediaries is of importance as they essentially function as the gatekeepers for access to water. Consumers have no control over the tap that delivers the water in the kiosks and the yards where they are located. As they depend on the intermediary to provide access to these taps, they have to adapt and adjust to the agenda of the intermediary in order to access water.

The role played by the intermediaries in our case studies also highlights the importance of how technologies are being used and implemented. In principle, there is no reason why the intermediary could not adhere to the guidelines forwarded by the WSP and WASREB. The intermediary could adhere to the price of 1.5-2 Ksh. per 20 litres and ensure that the tap can be accessed for a period of at least 12 hours per day. However, the everyday practices of using these technologies in the cases studied deviate significantly from this guideline. As a result, a situation the currently service delivery system appears to benefit the water service provider and the intermediary more than the consumers. In other words, how a particular technology is used and implemented in the everyday determines to a large extent for which actor that technology is truly appropriate. In the three cases studied in this article, it appears that the appropriate technologies are particularly appropriate for the water service provider and the intermediaries.

.

7

CONCLUSION

7.1 INTRODUCTION

In this final chapter, I use the results of the preceding chapters to reflect on how and why commercialised urban water utilities mobilise and implement pro-poor interventions. I started this discussion in chapter 2 with a literature review, to find out how the concept of pro-poor services is understood and used within water utilities. I followed this up to show empirically in chapters 3, 4 and 5 why the utilities adopt these strategies to serve the LIAs and how they are actually implemented. Chapter 6 focused on the impacts of the interventions for the LIA consumers. The findings in Chapters 3, 4, 5 and 6, as well as evidence from the literature review in Chapter 2 showed that utilities rather pragmatically adopt pro-poor strategies, partly using them to protect themselves from the financial and operational risks of service provision to low-income areas, while still satisfying the mandate of universal access. An important question then is how the interests of consumers are served. The name pro-poor water services suggest the implementation of measures and interventions that will improve access to water supply for 'poor' consumers in low-income areas. My findings and especially Chapter 6 shed doubts about whether this indeed happens. Although pro-poor projects and interventions have allowed utilities to expand coverage in the low-income areas, access to water remains difficult and costly for consumers living in these areas. These findings resonate with what critics of the concept have argued: they have raised questions about the equity implications of pro-poor strategies. In spite of such critiques, the concept remains popular, not only among governments and donor organisations but also among the managers of the utilities at the receiving end of such critiques. I begin the chapter with a summary on why utilities choose to serve low-income areas through pro-poor strategies and end by reflecting on the issues of equity raised around the concept.

7.2 PRO-POOR AS A PRAGMATIC MOVE TO BALANCE DUAL MANDATES OF WATER UTILITIES

The three utilities studied, like most other commercial utilities, have two mandates: the mandate to recover the cost associated with their operations and the mandate to ensure universal access. Yet, the performance benchmarks and indicators used to assess their activities focus more on cost recovery and financial efficiencies than on ensuring access to all. In Kenya for instance, six out of the nine key performance indicators (KPIs) used in the assessment of the service providers focus on the financial viability or sustainability of the utilities. It is only recently (in 2019) that the regulator presented guidelines for service provisioning to the un/underserved communities. The focus on financial indicators pushes utilities to mainly focus on extending services to locations where they are sure of being able to recover cost. As discussed throughout the chapters of this thesis, this makes utilities often reluctant to invest their limited resources in the low-income areas in which the risks of unpaid water bills and vandalism of water infrastructure are high. However, the mandate to ensure universal access makes that the utilities do also need to direct attention to these low-income areas, where often a larger proportion of the population they are supposed to serve reside. The thesis shows that pro-poor strategies have become a favoured choice of utility managers to both ensure financial sustainability while also meeting goals of universal access in cities with distinct well planned and unplanned settlements. This is so, because pro-poor strategies reduce the costs and risks of providing water services to low-income areas. Hence, for utility managers, pro-poor strategies are not so much, or not just, about tailoring the provision or services to the needs of a specific consumer segment, but also importantly tools to reduce risks. As shown in chapters 4 and 6, pro-poor strategies allow the utilities to delegate the last mile of water supply to the LIAs to intermediaries, who become responsible for collecting payments for their services. Besides, the use of appropriate technologies to extend services also means limiting the investment cost for the utilities. This because the cost involved in putting up these technologies (with the exception of the PPDs) is lower in comparison with the conventional connections and also because such interventions are often sponsored by donor organisations. In all, pro-poor water strategies allow the utility to protect itself from the financial and operational risks of service provision to low-income areas while still satisfying the mandate of universal access (Chapter 2).

7.3 PRO-POOR STRATEGIES AS A PRAGMATIC MOVE TO ALIGN THE INTEREST OF STAKEHOLDERS

Another way in which this thesis has explained the use of pro-poor strategies for supplying water to LIAs is that they help different stakeholders or actors operating in the water supply sector to achieve their various goals. The different chapters in this thesis have shown how utilities operate in a network of actors - governments, donor organisations and informal water operators – who all depend on each other to achieve their different goals. Water utilities are the main vehicle for ensuring that governments meet the political commitment of universal access to water; in this sense the good performance of water utilities is important for maintaining popular trust in the government, and also important in terms of electoral politics. A good portion of the budget needed for the development and expansion of water infrastructure is provided by donor organisations. Just like governments, donors need to justify their behaviour and expenditures to their constituencies. In water supply, an important and straightforward way of doing this is by showing how many additional people are provided with access through their projects and interventions. Again, water utilities are the instruments of donors to realize such increases in coverage. For meeting water service delivery targets in low-income areas, water utilities also partner with and depend on the informal providers already operating here. The informal providers in turn need partnerships with utilities to legitimise their operations in the LIAs.

As shown in chapters 4 and 5, pro-poor strategies serve as a pragmatic device mobilized by all actors to reach their goals. Hence, utilities engage with informal operators even when the evidence points out that their role as intermediaries may compromise the quality and equity of services. It seems the utilities tolerate intermediaries since the partnerships with them present them with a low-risk and low-cost way to provide at least "some level of access" in the LIAs, allowing them to recover their cost without being physically present there. The use of such intermediaries in Kisumu even when the utility had replaced the post-paid water kiosks with pre-paid dispensers (PPDs) goes to show the level of dependence the utilities have on the on such intermediaries.

For utilities, pro-poor strategies are a sub-optimal compromise as their goal remains to provide the same services for all the consumers within their service areas. This would imply the use of in-house connections also for low-income consumers. The exploitative nature of the intermediaries presents a strong argument for them to work towards this goal. Chapter 5 indicates that the utilities are turning to the use of PPDs to either control or eliminate the intermediaries who limit access and make profit on the backs of the consumers. These PPDs are financed by donors who have an interest to reach more people; an interest which can be duly served by investing in the PPDs rather than in individual household connections.

7.4 REFLECTIONS: THE EQUITY OF PRO-POOR SERVICES

Development practitioners promoting pro-poor services put a lot of emphasises on their ability to improve access for consumers living in the LIAs. There is ample literature to show that such strategies improve equity and better align water services to the specific needs and demands of the consumers in the low-income areas (Berg and Mugisha, 2010; Cross and Morel, 2005; Kayaga and Franceys, 2007; Mara, 2003; Sansom et al., 2004, Nhema and Zinyama, 2016; Watson et al., 1997; World Bank, 2000; Younger, 2007).

However, results from this thesis show that while the use of pro-poor strategies has led to improvements in some aspects of access such as coverage, other important aspects of access including affordability and service reliability remain questionable. In assessing the success of pro-poor strategies within a policy context, that emphasizes cost recovery and the commercialization of utilities, policy practitioners and donors often measure achievements against ideals, rather than examining in detail how projects are implemented and with what effects. For instance, the pro-poor technologies (including water kiosks and yard taps) used by Kenyan water utilities are estimated to serve at least 200 people. By financing the construction of such projects donors and development practitioners report 'improvements' in access for this number of people in the LIAs. However, in practice, true improvements in access are only achieved when the supply of water to these water points by the utilities is regular and stable; when the continued presence of intermediaries (including kiosks operators and landlords) who manage the water points to serve consumers is guaranteed and when the cost of the water provided is reasonable. Results from this thesis show that water service provision in LIAs does not meet these conditions, resulting in a limited reliability and affordability. In addition to a limitation in availability because of bulk water rationing by the utilities, consumers also have to contend with the unavailability of the intermediaries to serve them when they need water. Further, the profit seeking behaviour of the intermediaries translates into higher water prices, further compromising affordability for the consumers. These findings show that measuring the success of the pro-poor concept needs to move beyond numbers of additional people serviced, and be prepared to revisit assumptions about how pro-poor water services provision happens on the basis of investigations of what actually happens in LIAs.

My findings support wider critiques on the ability of pro-poor strategies to ensure equity in access to services. Service differentiation intrinsically implies that low-income consumers are provided with a different type of service. I show that in practice this means a lower service level at a higher cost. Prevailing critiques are often directed at the managers of the water utilities, who are directly responsible for implementing pro-poor strategies. In my analysis, I have emphasized that water utilities do not operate in an institutional vacuum: they have to meet a number of sometimes contradictory demands from governments and donors, and have to navigate multiple relations of accountability that intersect in complex ways. For them, the use of pro-poor strategies to supply water to the LIAs provides a convenient way of doing this. In this sense, utilities are merely the vehicles for implementing a concept pushed onto them by financiers; a concept that often they have not invented or initiated. To say this differently, utilities cannot be or become solely responsible for ensuring that pro-poor strategies meet the goal of improving access for low-income consumers or ensure equity of access across consumers. It is important to realize that that they depend on and are conditioned by what others do, especially in terms of financial and institutional arrangements. Governments especially, owe it to utilities to help minimize some of the challenges and risks that providing services to LIAs entails. For donor organisations and financiers, instead of handing down concepts or choose which ones to supports, there is a need to better consider the different contexts within which the utilities operate. The emphasis on approaches that 'work' across the sector may not only result in them failing but also defeat the goal of improving equity/access for the less served groups.

REFERENCES

Abubakar, I. R. (2019). Factors influencing household access to drinking water in Nigeria. Utilities Policy, 58: 40-51.

Adank, M. and Tuffour, B. (2013). Management models for the provision of small town and peri-urban water services in Ghana. TPP synthesis report. Retrieved May 20, 2016, from: https://www.ircwash.org/sites/default/files/tpp_synthesis_report_final_medium_res_min_size.pdf

Alda-Vidal, C., Kooy, M. and Rusca, M. (2018). Mapping operation and maintenance: An everyday urbanism analysis of inequalities within piped water supply in Lilongwe, Malawi. Urban Geography, 39(1): 104-121.

Alford, J. and Hughes, O. (2008). Public value pragmatism as the next phase of public management. The American Review of Public Administration, 38: 130-148.

Almansi, F., Hardoy, A. and Hardoy, J. (2010). Improving Water and Sanitation provision in Buenos Aires: What can a Research-oriented NGO do? IIED Human Settlements Working Paper Series. Accessed June 2 2017, from: http://pubs.iied.org/pdfs/10583IIED.pdf?

Amenga-Etego, R. and Grusky, S. (2005). The new face of conditionalities: The World Bank and water privatization in Ghana. The age of commodity: Water privatization in Southern Africa, 275-292.

Angel, J. and Loftus, L. (2017). 'With-against-and-beyond the human right to water', Geoforum, 98: 206-213.

Appelblad Fredby, J. and Nilsson, D. (2012). From "All for some" to "Some for all"? A historical geography of pro-poor water provision in Kampala. Journal of Eastern African Studies, 7(1): 40-57.

Baietti, A., Kingdom, W. and Van Ginneken, M. (2006). Characteristics of well-performing public water utilities. Water Supply and Sanitation Working Notes Accessed 31 July 2019, from: http://documents.worldbank.org/curated/en/110951468155137465/pdf/372910WSS0Workingnote9.pdf.

Bakker, K. (2003). Archipelagos and networks: urbanization and water privatization in the South. Geographical Journal, 169 (4): 328-341.

Bakker, K. (2007). The "commons" versus the "commodity": Alter-globalization, anti-privatization and the human right to water in the global south. Antipode, 39(3): 430-455.

Bakker, K., Kooy, M., Shofiani, N.E. and Martijn, E.J. (2008). Governance failure: rethinking the institutional dimensions of urban water supply to poor households. World Development, 36 (10): 1891-1915.

Batley, R. (2006). Guest editor's preface. Symposium on non-state provision of basic services. Public Administration and Development 26 (3): 193–196.

Bayliss, K., and Fine, B. (2007). Privatization and alternative public sector reform in Sub-Saharan Africa: Delivering on electricity and water. London: Palgrave MacMillan

Bell, M. and Franceys, R. (1995). Improving Human Welfare through Appropriate Technology: Government Responsibility, Citizen Duty or Customer Choice. *Social* Science Medicine, 40(9):1169-1179.

Benjaminsen, T.A. (2002). Formalising land tenure in rural Africa. Forum for Development Studies 29(2): 362–366. Accessed 20 June 2016, from https://doi.org/10.1080/08039 410.2002.96662 12.

Berg, S. V. and Mugisha, S. (2010). Pro-poor water service strategies in developing countries: Promoting justice in Uganda's urban project. Water policy, 12(4): 589-601.

Boakye, A., Schwartz, K. and Zwarteveen, M. (under review). The Appropriateness of Appropriate Technologies in Improving Access to Water in Low-Income Areas: Evidence from Kenya. International Journal of Water Resources Development.

Boakye-Ansah, A. S., Ferrero, G., Rusca, M. and van der Zaag, P. (2016). Inequalities in microbial contamination of drinking water supplies in urban areas: The case of Lilongwe, Malawi. Journal of water and health, 14(5): 851-863.

Boakye-Ansah, A. S., Schwartz, K. and Zwarteveen, M. (2019). Unravelling pro-poor water services: what does it mean and why is it so popular? Journal of Water, Sanitation and Hygiene for Development, 9(2): 187-197.

Boakye-Ansah, A. S., Schwartz, K. and Zwarteveen, M. (2019). From rowdy cartels to organized ones? The transfer of power in urban water supply in Kenya. The European Journal of Development Research, 31(5): 1246-1262.

Bohman, A. (2012). The presence of the past: a retrospective view of the politics of urban water management in Accra, Ghana. Water History, 4(2): 137-154.

Boland John J. and Whittington D. (1998). The political economy of increasing block tariffs in developing countries: Special Paper Economy and Environment Programme for Southeast Asia. Accessed July 3 2016, from: https://idlbncidrc.dspacedirect.org/bitstream/handle/10625/31784/118117.pdf?seq uence=8&isAllowed=y

Bond, P. (2004). Water commodification and decommodification narratives: pricing and policy debates from Johannesburg to Kyoto to Cancun and back. Capitalism Nature Socialism, 15(1): 7-25.

Bond, P. and Dugard, J. (2008). 'The case of Johannesburg water: What really happened at the pre-paid 'Parish pump', Law, Democracy and Development, 12(1): 1-28. Nature Socialism, 15(1): 7–25

Brendel, D. (2006). Healing Psychiatry: Bridging the Human/Science Divide, Cambridge: MIT Press

Burt, Z. and I. Ray. (2014). Storage and non-payment: persistent informalities within the formal water supply of Hubli-Dharwad, India. Water Alternatives 7 (1): 106–120.

Carter, R., Tyrrel, S. and Howsam, P. (1999). The impact and sustainability of community water supply and sanitation programmes in developing countries. Water and Environmental Journal, 13(4): 292-296.

Castro, V., and Morel, A. (2008). Can delegated management help water utilities improve services to informal settlements? Waterlines, 27(4): 289-306.

Chakava, Y. and Franceys, R. (2016). What hope for the transition? Evaluating pro-poor water supply interventions in urban low-income settlements in Kenya. Waterlines, 35(4): 365-387.

Cleaver, F. (1999). Paradoxes of participation: Questioning participatory approaches to development. Journal of International Development, 11(4): 597-612.

Conan, H. (2003). Scope and scale of small scale independent private water providers in 8 Asian cities: preliminary findings. Asian Development Bank working paper, Manila.

Conan, H. and M. Paniagua. (2003). The role of small scale private water providers in serving the poor. Asian Development Bank working paper, Manila.

Connors, G. (2005). When utilities muddle through: Pro-poor governance in Bangalore's public water sector. Environment and Urbanization, 17(1): 201-218.

Cosgrove, W.J. and F.R. Rijsberman. (2000). World water vision: Making water everybody's business. London: Earthscan.

Crique, L. (2015). 'Infrastructure urbanism: Roadmaps for servicing unplanned urbanisation in emerging cities', Habitat International, 47: 93-102.

Cross, P. and Morel, A. (2005). Pro-poor strategies for urban water supply and sanitation services delivery in Africa. Water Science and Technology, 51(8): 51-57.

Dagdeviren, H. (2008). Waiting for miracles: The commercialization of urban water services in Zambia. Development and Change, 39(1): 101-121.

Dagdeviren H. and Robertson S. (2009). Access to water in the slums of the developing world. International Policy Centre for Inclusive Growth (IPC), Working Paper No. 57. Retreived 13 september 2016, from: (http://www.ipc-undp.org/pub/IPCWorkingPaper57.pdf)

De Albuquerque, C. and Winkler, I. T. (2010). Neither Friend nor Foe: Why the Commercialization of Water and Sanitation Services is Not the Main Issue in Realization of Human Rights. Brown J. World Aff., 17: 167.

De Soto, H. (2000). The mystery of capital: Why capitalism triumphs in the West and fails everywhere else. New York: Basic Civitas Books.

Deedat, H. and Cottle, E. (2002). 'Cost recovery and prepaid water meters and the cholera outbreak in KwaZulu-Natal: A case study in Madlebe' , In: McDonald, D. and Pape, J. (eds.), Cost Recovery

Dill, B. (2009). The Paradoxes of Community-based Participation in Dar es Salaam, Development and Change. 40(4): 717-743.

DiMaggio, P. and Powell, W. (1983). 'The iron cage revisited: Institutional isomorphism and collective rationality in organizational fields', American Sociological Review, 48(2): 147-160.

Ehrhardt, D., Groom, E., Halpern, J. and O'Connor, S. (2007). Economic Regulation of WSS services: Some Practical Lessons. Water Sector Board Discussion Paper #9, April 2007, Washington, DC: World Bank.

Feachem, R. (1980). Community Participation in Appropriate Water Supply and Sanitation technologies: The Mythology for the Decade, Proceedings of the Royal Society of London. Series B, Biological Sciences. 209(1174): 15-29.

Franceys, R. (2005). Charging to enter the water shop? The costs of urban water connections for the poor, Water Science and Technology: Water Supply. 5(6): 209-216.

Franceys, R. (2008). GATS, 'privatization' and institutional development for urban water provision: Future postponed'?. Progress in Development Studies, 8(1): 45-58.

Frediani, A. A. (2015). Space and capabilities: Approaching informal settlement upgrading through a capability perspective. In; The City in Urban Poverty (pp 64-84). Palgrave Macmillan, London.

Freire, M. and Stren, R. (2001). The challenge of urban government: policies and practices. The World Bank.

Furlong, K. (2010). Neoliberal water management: Trends, limitations, reformulations. Environment and Society: Advances in Research, 1(1): 46-75.

Furlong, K. (2014). STS beyond the "modern infrastructure ideal": Extending theory by engaging with infrastructure challenges in the South. Technology in Society 38: 139–147.

Furlong, K. (2015). Water and the entrepreneurial city: The territorial expansion of public utility companies from Colombia and the Netherlands. Geoforum, 58: 195-207.

Furlong, K. and Bakker, K. (2010). The contradictions in 'alternative'service delivery: governance, business models, and sustainability in municipal water supply. Environment and Planning C: Government and Policy, 28(2): 349-368.

Furlong, K. and Kooy, M. (2017). Worlding water supply: thinking beyond the network in Jakarta. International Journal of Urban and Regional Research, 41(6):888-903.

Furlong, K., Carré, M. and Acevedo Guerrero, T. (2017). Urban service provision: Insights from pragmatism and ethics. Environment and Planning A: Economy and Space, 49(12): 2800-2812.

Gandy M. (2004). Rethinking urban metabolism: water, space and the modern city, City, 8:3, 363-379, DOI: 10.1080/1360481042000313509

Gerlach E. and Franceys R. (2010). Regulating water services for all in developing economies. World Development, 38(9): 1229–1240.

Glaser, B. G. and Strauss, A. L. (1967). The discovery of grounded theory: Strategies for qualitative research: Transaction publishers.

Gleick, P. (1996). Basic water requirements for human activities: meeting basic needs. Water International, 21: 83-92.

Goff, M. and Crow, B. (2014). What is water equity? The unfortunate consequences of a global focus on 'drinking water'. Water international, 39(2): 159-171.

Gomez, J. and Nakat, A. (2002). Community Participation in Water and Sanitation. Water International, 27(3): 343-353.

Government of Kenya (1983). Five Year Development Plan, 1983–1988. Government Press, Kisumu

Government of Kenya (1990); Socio-Economic Profiles of the Ministry of Planning and Development and UNICEF

Government of Kenya (2002). The water Act 2002

Government of Kenya (2007a). The National Water Services Strategy (NWSS) (2007 – 2015).

Government of Kenya (2007b). The Pro-Poor Implementation Plan for Water Supply and Sanitation (PPIP - WSS).

Government of Kenya (2010). The Constitution of Kenya. Government of Kenya: Nairobi.

Government of Kenya 2016. The water Act 2016

Graham, S. and Marvin, S. (2001). Splintering urbanism: Networked infrastructures, technological mobilities and the urban condition. Psychology Press.

Gulyani, S., Talukdar, D. and Kariuki, R. M. (2005). Universal (non) service? Water markets, household demand and the poor in urban Kenya. Urban Studies, 42(8): 1247-1274.

Gunnerson, C.G. 1ulius, D. and Kalbermatten, 1. (1978). Alternative Approaches to SanitationTechnologies. Paper Presented at the Workshop on Water Pollution Problems, held by the International Association for Water Pollution Research, Stockholm, 14 June.

Gupta, J., Ahlers, R. and Ahmeed, L. (2010). 'The human right to water: Moving towards consensus in a fragmented world'. Review of European Community and International Environmental Law, 19(3): 294-305.

Hadzovic, L. (2014). Water Pricing Regimes and Production of the Urban Waterscape. A Case Study of Lilongwe, Malawi. Master's thesis UNESCO-IHE, Delft, ES.14.03

Hailu, D., Rendtorff-Smith, S. and Tsukada, R. (2011). Small-Scale Water Providers in Kenya: Pioneers or predators? New York: United Nations Development Programme.

Harris, C. and Janssens, J. (2004). Operational guidance for World Bank Group staff: public and private sector roles in water supply and sanitation services (No. 43102, pp. 1-32). The World Bank. Accessed 10 July 2017. From;

http://documents.worldbank.org/curated/en/556271468155727649/pdf/431020WP
01NO0P1vateSector1Role s1WSS .pdf.

Harvey, E. (2005). Managing the poor by remote control: Johannesburg's experiments with prepaid water meters. The age of commodity: Water privatization in Southern Africa, 120-127.

Harvey, P. and Reed, R. (2007). Community-managed water supplies in Africa: sustainable or dispensable? Community Development Journal, 42(3): 365-378.

Heymans, C., Eales, K. and Franceys, R. (2014). The limits and possibilities of prepaid water in urban Africa: Lessons from the field. Accessed 14 May 2016. From: http://wwwwds.worldbank.org/external/default/WDSContentServer/WDSP/IB/2 014/08/26/000470435_20140826140705/Rendered/PDF/901590REPLACEM0P repaid0Water0Africa.pdf

Hukka, J. J. and Katko, I. (2003). Refuting the paradigm of water services privatization. Natural Resource Forum, 27: 142–155.

Hunter, P, MacDonald, A. and Carter, R. (2010). Water Supply and Health. PLoS Medicine, 7(11): e1000361.

Jaglin, S. (2008). Differentiating networked services in Cape Town: Echoes of splintering urbanism? Geoforum, 39(6): 1897-1906.

Jaglin, S. (2014). Regulating service delivery in Southern cities: rethinking urban heterogeneity. In, The Routledge handbook on cities of the Global south. Abingdon: Routledge, 456–469.

Jaglin, S. (2016). Is the network challenged by the pragmatic turn in African cities? Urban transition and hybrid delivery configurations. In; Beyond the Networked City. Infrastructure reconfigurations and urban change in the North and South. Abingdon: Routledge, 182-203

Jaglin, S. (2002). The right to water versus cost recovery: participation, urban water supply and the poor in sub-Saharan Africa. Environment and Urbanization, 14(1): 231-245.

Jimenez-Redal, R., Parker, A. and Jeffrey, P. (2014). Factors influencing the uptake of household water connections in peri-urban Maputo, Mozambique. Utilities Policy, 28: 22-27

Johnstone, N., Wood, L. and Hearne, R. R. (1999). The regulation of private sector participation in urban water supply and sanitation: realising social and environmental objectives in developing countries. Environmental Economics Programme, No. 1032-2016-83914.

Kariuki, M. and Schwartz, K. (2005). Small-scale private service providers of water supply and electricity: A review of incidence, structure, pricing, and operating characteristics. The World Bank.

Kayaga, S. and Franceys, R. (2007). Costs of urban utility water connections: Excessive burden to the poor'. Utilities Policy, 15: 270-277.

Kemendi, T.J. and Tutusaus, M. (2018). The impact of pro-poor interventions on the performance indicators of a water utility: case studies of Nakuru and Kisumu. Journal of Water, Sanitation and Hygiene for Development, 8 (2): 208-216.

Kericho Water and Sanitation Company (KEWASCO). (2018). Pro-poor strategy for Kericho's Informal Settlements (Low Income Areas) January 2018 – June 2022

Kessides, I. (2004). Reforming Infrastructure Privatization, Regulation, and Competition. The World Bank.

Kirkpatrick, C. and Parker, D. (2005). Domestic regulation and the WTO: The case of water services in developing countries. World Economy, 28(10): 1491-1508.

KIWASCO (2018). Field Report on Prepaid Dispenser Operations, Kisumu: KIWASCO.

KIWASCO (2018). Field Report on Prepaid Dispensers Operations, Internal Report, Kisumu: KIWASCO.

Kjellén, M. (2000). Complementary water systems in Dar es Salaam, Tanzania: the case of water vending. International Journal of Water Resources Development, 16(1): 143-154.

Klein, M. U. (1996). Economic regulation of water companies. World Bank Policy Research Working Paper, 1649.

Komives, K. (1999). Designing Pro-Poor Water and Sewer Concessions: Early Lessons from Bolivia, Policy Research Working Paper 2243. Washington D.C.: World Bank.

Laugeri, L. (1986). Institutional Development in Community Water Supply and Sanitation. WHO Publication.

Lauria, D. T., Hopkins, O. S. and Debomy, S. (2005). Pro-poor subsidies for water connections in West Africa. Water Supply and Sanitation Working Note, 3.

Lobina, E. (2005) 'Problems with Private Water Concessions: A Review of Experiences and Analysis of Dynamics', Water Resources Development 21(1): 55– 87.

Loftus, A. (2006). Reification and the dictatorship of the water meter. Antipode, 38(5): 1023-1045.

MacGrannahan, G. and Sattherwaite (2006). Governance and Getting the Private Sector to Provide Better Water and Sanitation Services to the Urban Poor. IIED Human Settlements Discussion Paper Series. Accessed 23 March 2017. From; https://www.ircwash.org/sites/default/files/McGranahan-2006-Governance.pdf.

MacIntosh, A. (2003). Asian Water Supplies: Reaching the Poor. Manila: Asian Development Bank.

Manor, J. (2004). User committees: A potentially damaging second wave of decentralisation? The European Journal of Development Research, 16(1): 192-213.

Mara, D. (2003). Water, Sanitation and Hygiene for the health of developing Nations. Public Health, 117: 452-456.

Mara, D. and Alabaster, G. (2008). A new paradigm for low-cost urban water supplies and sanitation in developing countries. Water Policy, 10(2): 119.

Marson, M. and Savin, I. (2015). Ensuring sustainable access to drinking water in sub saharan africa: conflict between financial and social objectives. World Development, 76: 26-39.

Mason, N., Doczi, J. and Cummings, C. (2016). Innovating for pro-poor services: Why politics matter', ODI Insights. Accessed; 20 February 2017. From; https://www.odi.org/sites/odi.org.uk/files/resource-documents/10349.pdf

Martin, P. Y. and Turner, B. A. (1986). Grounded theory and organizational research. The journal of applied behavioural science, 22(2), 141-157.

Mbuvi, D. Dorcas, M. and Schwartz, K. (2013). 'The politics of utility reform: A case study of the Ugandan water sector'. Public Money and Management, 33(5): 377-382.

McDonald, D. and Ruiters, G. (2005). The age of commodity: Privatizing water in Southern Africa. London: Earthscan

McDonald, D.A. (2002). The theory and practice of cost recovery in South Africa. In; cost recovery and the crisis of service Delivery in South Africa. London: Zed Books; Cape Town: HSRC Publishers, 17–40.

Ministry of Foreign Affairs of the Netherlands (2016). Contributing to water, sanitation and hygiene for all, forever. DGIS/IGG draft WASH policy and strategy note 2016–2030.

Mirosa, O. and Harris, L. (2011). Human right to water: Contemporary challenges and contours of a global debate. Antipode, 44(3): 932-949.

Misra, K. (2014). From formal-fnformal to emergent formalisation: fluidities in the production of urban watercapes. Water Alternatives 7 (1): 15–34.

Moore, M. (1994). Public value as the focus of strategy. Australian Journal of Public Administration, 53(3): 296-303.

Moretto, L. (2007). Urban governance and multilateral aid organizations: The case of informal water supply systems. The Review of International Organizations, 2(4): 345-370.

Mumma, A. (2007). Kenya's New Water Law: An Analysis of the Implications of Kenya's Water Act, 2002, for the Rural Poor. In; Community-based water law and water resource management reform in developing countries, 158–172.

Nakuru Water and Sanitation Company (NAWASSCO) 2013. "Strategies and Actions towards Improving Water and Sanitation Service Provision to Nakuru's Low income Areas". Nakuru Water and Sanitation Service Company Pro-poor Strategy, January 2013 - December 2017

Narsiah, S. (2010). The Neoliberalisation of the Local State in Durban, South Africa. Antipode, 42(2): 374–403.

National Water Master Plan (2013). The Project on the Development of the National Water Master Plan 2030, https://wasreb.go.ke/national-water-master-plan-2030.

Njiru, C. (2004). Utility-small water enterprise partnerships: serving informal urban settlements in Africa. Water Policy, 6(5): 443-452.

Njiru, C., Smout, I. and Sansom, K. (2001). Managing water services through service differentiation and pricing in an African city. Water and Environment Journal, 15(4), 277-281.

Octrinsic Technologies (2017). Diagnosis and Repair of Prepaid Water Meters, Internal Report, Nakuru: NAWASSCO.

Onjala, J. O. (2002). Managing water scarcity in Kenya: Industrial response to tariffs and regulatory enforcement. (Doctoral dissertation, University of Nairobi).

O'Rourke, E. (1992). The international drinking water supply and sanitation decade: dogmatic means to a debatable end. Water Science and Technology, 26(7-8): 1929-1939.

Owuor, S. O. and Foeken, D. (2012). From self-help group to Water Company: The Wandiege Community Water Supply Project (Kisumu, Kenya). In; Transforming innovations in Africa (pp. 127-147). Brill.

Paterson, C. Mara, D. and Curtis, T. (2007). Pro-poor sanitation technologies. Geoforum, 38: 901-907.

Peters, K., and Oldfield, S. (2005). The paradox of 'free basic water'and cost recovery in Grabouw: Increasing household debt and municipal financial loss. In Urban Forum (Vol. 16, No. 4, pp. 313-335). Springer Netherlands.

Pihljak, L. H., Rusca, M., Alda-Vidal, C. and Schwartz, K. (2019). Everyday practices in the production of uneven water pricing regimes in Lilongwe, Malawi. Environment and Planning C: Politics and Space, 2399654419856021

Prasad, N. (2006). Privatization Results: Private Sector Participation in Water Services after 15 Years, Development Policy Review 24(6): 669–92

Ranganathan, M. (2014). 'Mafias' in the waterscape: Urban informality and everyday public authority in Bangalore. Water Alternatives 7 (1): 89–105.

Renouf, R. (2016). Pro-poor tariff systems: What if 'fair'pricing isn't enough? Accessed 12 January, 2016, from: http://www.wsup.com/2016/05/16/pro-poor-tariff-systems/

Rusca, M. and Schwartz, K. (2012). Divergent sources of legitimacy: A case study of international NGOs in the water services sector in Lilongwe and Maputo. Journal of Southern African Studies, 38(3): 681-697.

Rusca, M. and Schwartz, K. (2018). The paradox of cost recovery in heterogeneous municipal water supply systems: ensuring inclusiveness or exacerbating inequalities? Habitat International, 73: 101-108.

Rusca, M., Boakye-Ansah, A.S., Loftus, A., Ferrero, G. and van der Zaag, P. (2017). An interdisciplinary political ecology of drinking water quality. Exploring socio-ecological inequalities in Lilongwe's water supply network. Geoforum, 84: 138-146.

Rusca, M., Schwartz, K., Hadzovic, L. and Ahlers, R. (2015). Adapting generic models through bricolage: Elite capture of water users' associations in peri-urban Lilongwe. European Journal of Development Research, 27(5): 777-792.

Ryan, P. and Adank M. (2010). Mechanisms to Ensure Pro-Poor Water Service Delivery in Peri-Urban and Urban areas. TPP Working Document. Accessed 20 May 2016, from http://www.ircwash.org/sites/default/files/Ryan-2010-Mechanisms.pdf

Savelli, E., Schwartz, K. and Ahlers, R. (2018). The Dutch aid and trade policy: Policy discourses versus development practices in the Kenyan water and sanitation sector. Environment and Planning C: Politics and Space, 0263774X18803364.

Schaub-Jones, D. (2008). Harnessing entrepreneurship in the water sector: Expanding water services through independent network operators. Waterlines 27 (4): 270–288.

Von Schnitzler, A. (2008). Citizenship prepaid: Water, calculability, and techno-politics in South Africa. Journal of Southern African Studies, 34(4): 899-917.

Schumacher, E. F. (1973). Small is beautiful: A study of economics as if people mattered. London: Blond and Briggs.

Schwartz, K. (2006). Managing public water utilities: An assessment of bureaucratic and new public management models in the water supply and sanitation sectors in low-and middle-income countries. PhD, Erasmus University Rotterdam

Schwartz, K. (2008). The New Public Management: The future for reforms in the African water supply and sanitation sector? Utilities Policy, 16: 49-58.

Schwartz, K. and Sanga, A. (2010). Partnerships between utilities and small-scale providers: Delegated management in Kisumu, Kenya. Physics and Chemistry of the Earth, Parts A/B/C, 35(13): 765-771.

Schwartz, K. Boakye-Ansah, A.S. and Tutusaus M. (2020). Experiences with Pre-Paid Dispensers. Lessons from three Kenyan cities. Working paper for the project Boosting Effectiveness of Water Operators' Partnerships (BEWOP)

Schwartz, K., Tutusaus Luque, M., Rusca, M. and Ahlers, R. (2015). (In) formality: the meshwork of water service provisioning. Wiley Interdisciplinary Reviews: Water, 2 (1): 31-36.

Schwartz, K., Tutusaus, M. and Savelli, E. (2017). Water for the urban poor: balancing financial and social objectives through service differentiation in the Kenyan water sector. Utilities Policy: 48, 22-31.

Shields, P. M. (2008). Rediscovering the taproot: Is classical pragmatism the route to renew public administration? Public Administration Review, 68(2): 205-221.

Shirley, M. M. (1999). Bureaucrats in business: The roles of privatization versus corporatization in state-owned enterprise reform. World development, 27(1): 115-136.

Sima, L. and Elimelech, M. (2011). The informal small-scale water services in developing countries: The business of water for those without formal municipal connections. New Haven: Wiley-Blackwell, 231

Sjaastad, E. and B. Cousins. 2009. Formalisation of land rights in the South: An overview. Land Use Policy, 26 (1): 1–9.

Smith, L. (2004). The murky waters of the second wave of neoliberalism: corporatization as a service delivery model in Cape Town. Geoforum, 35(3): 375-393.

Smith, L. (2006). 'Neither Public Nor Private: Unpacking the Johannesburg Water Corporatization Model'. Social Policy and Development Programme Paper No 27. Geneva: United Nations Research Institute for Social Development

Smith, L. and Hanson, S. (2003). Access to water for the urban poor in Cape Town: where equity meets cost recovery. Urban Studies, 40(8): 1517-1548.

Snell, S. M. (1998). Water and sanitation services for the urban poor: small-scale providers-typology and profiles (No. 33106, pp. 1-59). The World Bank.

Solo, T., Perez, E. and Joyce, S. (1993). Constraints in Providing Water and Sanitation Services to the Urban Poor. WASH Technical Report 85. Washington D.C.: USAID.

Stoler, J., Weeks, J. R. and Otoo, R. A. (2013). Drinking water in transition: a multilevel cross-sectional analysis of sachet water consumption in Accra. PLoS One, 8(6): e67257.

Tewari, D. D. and Shah, T. (2003). An assessment of South African prepaid electricity experiment, lessons learned, and their policy implications for developing countries. Energy policy, 31(9): 911-927.

Tiwale, S., Rusca, M. and Zwarteveen, M. (2018). The power of pipes: Mapping urban water inequities through the material properties of networked water infrastructures-The case of Lilongwe, Malawi. Water Alternatives, 11(2): 314-335.

Trémolet, S. and Hunt, C. (2006). Taking account of the poor in water sector regulation. Water Supply and Sanitation Working Notes, 11.

Tutusaus, M. (2019). Compliance or Defiance? Assessing the Implementation of Policy Prescriptions for Commercialization by Water Operators, Leiden: CRC Press/Balkema.

Tutusaus, M., Cardoso, P. and Vonk, J. (2018). (de) Constructing the conditions for private sector involvement in small towns' water supply systems in Mozambique: policy implications. Water Policy, 20(S1): 36-51.

UNECE and WHO (2013). The Equitable Access Score-card: Supporting policy processes to achieve the human right to water and sanitation. Geneva. Retrieved 17 January 2019, from: http://www.unece.org/fileadmin/DAM/env/water/publications/PWH_equitable_a ccess/1324456_ECE_MP_WP_8_Web_Interactif_ENG.pdf

UN-Habitat (2006). State of the World's Cities 2006/7, London: Earthscan.

UN-Habitat (2016). Enhancing for pro-poor Water, Sanitation and Hygiene (WASH) governance through improved decision-making and performance management in Cambodia: Report on Assessment of gaps and needs for institutional strengthening of sub-national levels in WASH sector in Cambodia. Retrieved 21

December 2018, from:
https://washenablingenvironment.files.wordpress.com/2017/03/output-
1_assessment-of-gaps-and-needs-on-wash-capacity-in-cmb.pdf

UN-HABITAT (2013). Country programme document, 2013-2015, Kenya. Regional
urban study profile. http://unhabitat.org/books/kenya-national-urban-profile/
United Nations Human Settlements Programme

UNICEF and WHO. (2015). Progress on sanitation and drinking water – 2015 update and
MDG assessment (Report). New York: UNICEF and World Health Organization,
90 pp

United Nations (2010). Resolution Adopted by the General Assembly on July 2010. The
Human Right to Water and Sanitation. UN.
http://www.un.org/waterforlifedecade/human_right_to_water.shtm

Van Rooyen, C. and Hall, D. (2007). Public is as private does: The confused case of Rand
Water in South Africa: Municipal Services Project.

Van Ryneveld, P., Parnell, S. and Muller, D. (2003). Indigent policy: Including the poor
in the city of Cape Town's income strategy. Cape Town: City of Cape Town.

Water and Sanitation for the Urban poor [WSUP] (2015). Stand-alone Unit or
Mainstreamed Responsibility: How can Water Utilities Serve Low-income
Communities? Accessed on 22 February 2017, from: http://www.wsup.com/wp-
content/uploads/2015/02/DP005-Stand-alone-unit-or-mainstreamed-
responsibility.pdf

Water and Sanitation Program [WSP] (2005). Rogues no more? Water kiosk operators
achieve credibility in Kibera. Water and sanitation program field note. Accessed on
18 March 2016 from http://www-
wds.worldbank.org/external/default/WDSContentServer/WDSP/IB/2005/07/22/00
0011823_20050722175330/Rendered/PDF/330630PAPER0Rogues120No120Mo
re10Kibera.pdf.

Water and Sanitation Program [WSP] (2009b). Improving water utility services through
delegated management. Lessons from the utility and small-scale providers in
Kisumu, Kenya. Accessed on 22 April 2016 from, https://www.wsp.org/sites
/wsp.org/files /publications/Af-imp_through_delegated_mgmt.pdf.

Water and Sanitation programme [WSP] (2008). Performance Improvement Planning:
Upgrading and Improving Urban Water Services, 44125, Washington D.C.: World
Bank.

Water and Sanitation Programme [WSP] (2009a). Setting up pro-poor units to improve
service delivery: lessons from water utilities in Kenya, Tanzania, Uganda and
Zambia. Water and sanitation program field note. Accessed on 5 March 2016,
from:
http://documents.worldbank.org/curated/en/958701468041384220/pdf/521990W
SP0serv10Box345554B01PUBLIC1.pdf

Water Service Trust Fund [WSTF] (2010). Formalising water supply through partnerships. The Mathare-Kosovo water model. Accessed on 12 April 2017 from, http://www.waterfund.go.ke/toolkit/Downloads/2.%20WSPs%20Supplying%20In formal %20Settleme nts.pdf.

Water Services Regulatory Board [WASREB] (2014). A performance review of Kenya's Water Services Sector 2014–2015. Issue No. 7/2014

Water Services Regulatory Board [WASREB] (2015). A performance review of Kenya's Water Services Sector 2014–2015. Issue No. 8/2015

Water Services Regulatory Board [WASREB] (2016). A performance review of Kenya's Water Services Sector 2014–2015. Issue No. 9/2016.

Water Services Regulatory Board [WASREB] (2019). A performance review of Kenya's Water Services Sector 2014–2015. Issue No. 11/2019

Water Utility Partnership [WUP] (2003). Better Water and Sanitation for the Urban Poor: Good Practice from Sub-Saharan Africa. Accessed on 2 June 2017, from: https://openknowledge.worldbank.org/bitstream/handle/10986/14746/431010WP0 1NO0P1haran1WSS1Urban1Poor.pdf;sequence=1

Watson, G., Vijay, N., Gelting, R. and Beteta, H. (1997). Water and sanitation associations: Review and best practices. In A. Subramanian, N. Jagannathan, and R. Meinzen-Dick (Eds), User Organizations for Sustainable Water Services, World Bank Technical Paper (354).

Whittington, D. (1992). Possible Adverse Effects of Increasing Block Water Tariffs in Developing Countries. Economic Development and Cultural Change, 41(1): 75-87.

WHO and UNICEF (2011). Drinking Water Equity, safety and sustainability: JMP Thematic Report on Drinking Water 2011. Accessed on 18 February 2019, from: https://washdata.org/sites/default/files/documents/reports/2017-07/JMP-2011-tr-drinking-water.pdf

WHO and UNICEF Joint Monitoring Programme for Water Supply and Sanitation [JMP] (2017). Safely managed drinking water - thematic report on drinking water 2017. Geneva, Switzerland: World Health Organization; 2017. Licence: CC BY-NC-SA 3.0 IGO.

WHO and UNICEF. (2015). Progress on sanitation and drinking water: 2015 update and MDG assessment. Accessed on 23 April 2016 from: http://www.unicef.org/publications/files/Progress_on_Sanitation_and_Drinking_Water_2015_Update_.pdf

WHO (2003). The Right to Water. http://www2.ohchr.org/english/issues/water/docs/Right_to_Water.pdf

Wilder, M. and Romero Lankao, P. (2006). Paradoxes of decentralization: Water reform and social implications in Mexico. World Development, 34(11): 1977–1995

Winpenny, J. (2003). Financing water for all. Report of the world panel on financing water infrastructure. Accessed on 8 July 2017 from; http://www.oecd.org/green growth/21556 665.pdf.

World Bank (1992). Water Supply and Sanitation Projects: The Bank's Experience - 1967-1989. The World Bank. Accessed on 21 May 2016 from, http://www-wds.worldbank.org/external/default/WDSContentServer/WDSP/IB/1992/06/19/000009265_3961003042925/Rendered/PDF/multi_page.pdf

World Bank (2015). Leveraging water global practice knowledge and lending: Improving services for the Nairobi water and sewerage utility to reach the urban poor in Kenya. International Bank for reconstruction and development, and the World Bank.

APPENDICES

Appendix A: Interviewing guide

1. What are the existing technologies/infrastructure, financial and management/organisational mechanisms involved in water delivery in cities

2. Are all neighbourhoods served through/by the same mechanisms?

3. Why are the different neighbourhoods served differently: what informed the decision?

4. How was the different options/mechanisms introduced:

   - Were different options presented?

   - Why were specific interventions chosen over others?

5. How did the concept of pro-poor water delivery evolve in Kenyan water sector?

6. Who/what is behind the promotion of this concept: what informed the decision?

   - What criteria is used in the identification of the [urban] poor?

   - Who is involved in the identification process?

   - How is the process actually implemented (on the ground). Does the process cover/capture the target group?

   - What does this mean for the selection/implementation of pro-poor interventions: does it promote the selection of the right interventions/vice versa?

7. What are the laws, policies and regulations governing water delivery in Kenya?

   - Does these laws, policies and regulations capture the [urban] poor?

   - How did the present regulation governing water delivery come into force?

   - Are there specific laws and regulations targeting the urban poor?

8. How does these laws, policies/regulations influence decisions concerning pro-poor activities?

   - Does these laws, policies regulations capture the contributions of other practitioners in the sectors?

9. Who is involved in the implementation/monitoring of the pro-poor activities (laws, policies, regulations)?

- How are these actors identified?

- What role does such actors play in the process?

- How do these actors fulfil the roles assigned them?

- What are the outcomes/consequences on how these roles are fulfilled?

- What does this mean for the selection/implementation of pro-poor interventions: does it promote the selection of the right interventions/vice versa: how?

10. How do the other practitioners in the sector contribute to/influence (if any) the selection/implementation process

Appendix B: Data collection guide for the utilities

1. Quality of service
   a) Water coverage: the number of people served with drinking water by a utility expressed as a percentage of the total population within the service area of a utility
   b) Hours of supply: the average number of hours per day that a utility provides water to customers in LIAs
2. Financial sustainability
   a) Operations and maintenance cost: amount of monies spent on operations and maintenance in the selected LIAs
   b) Collection efficiency: the total amount collected by a utility expressed as a percentage of the total amount billed in a given period
3. Operational sustainability
   a) Staff productivity: number of staffs per 1000 connections
   b) NRW: difference between the amount of water produced and the amount billed
   c) Metering ratio: number of connections with functional meters

Appendix C: Questionnaire for household survey

Name of LIA:

County:

Population:

Purpose: To find out how the adoption of the pro-poor strategies is benefitting or otherwise the urban low-income neighbourhoods.

Part 1: Previous situation before the interventions: this information was to help establish a baseline for assessing the current situation

Family size

What was the number of people in your household?

Income spent on water

a)   How often were you paid for your work you do?

[1] Daily [2] Weekly [3] Monthly [4] Other: Specify

b)   How much did you earn

[1] Ksh ≤ 100 [2]101–500 [3] Ksh501–1000 [4] Ksh1001–1500 [5] Ksh1501–2000

c)   How often did you pay for water?

[1] Daily [2] Weekly [3] Monthly [4] Others: Specify

d)   How much did you pay for water? __Ksh

e)   How will you describe the amount you were spending on water?

[1] Low [2] Normal [3] High [4] Other: Specify

f)   How will you compare the amount paid for water with electricity?

Water availability: (skip g and h if response for f is NO)

a)   From which infrastructure did you collect water for this household?

[1] Water kiosk [2] Yard tap [3] Personal connection [4] Other: Specify

b)   How much water were you collecting in a day from this water source? ____Litres

c)   Was this water enough for your activities? [1] Yes [2] No

d)   What household activities did you use water from this source for?

[1] Drinking [2] Cooking [3] Laundering [4] Bathing [5] Toilet flushing [6] Other: Specify

e)   How much water did you need for daily activities? ___Litres

f)   Were you depending on other water sources or providers to meet your daily water use? [1] Yes [2] No

g)   Which were the other sources you depended on?

[1] Groundwater [2] Surface water [3] Rain water [4] Cart pushers [5] Other: Specify

h)   What activities did you use water from these other sources and providers for?

[1] Drinking [2] Cooking [3] Laundering [4] Bathing [5] Flushing toilet [5] other: Specify

Time spent collecting water

a)   How far did you have to walk to collect water?

b)   How long did you have to wait at the water source to collect water?

c)   How many trips did you make to the water point in a day?

Continuity of supply

a)   How many days in a week did you receive water?

b)   How many hours in a day did you receive water?

c)   Did you experience supply interruptions at your water point? [1] Yes [2] No

d)   How often did you experience these interruptions at your water point?

[1] Not often [2] Often [3] Very often [4] Others: Specify

e)   What is the response to maintenance whenever it is needed? [1] Prompt

[2] Delayed [3] If delayed Specify

g)   Which of the aspects of your water supply did you suggest need improvement?

[1] Pressure [2] Reliability [3] Maintenance [4] Cost [5] None [6] Other: Specify

h)   How would you rate water supply service prior to the interventions?

[1] Very Good [2] Good [3] Average [4] Poor [5] Other: Specify

Part 2: Current situation after the interventions were implemented

Family size

What is the number of people in your household?

 Income spent on water

a)   How often are you paid for the work you do

[1] Daily 2] Weekly [3] Monthly [4] Other: Specify

b)   How much do you earn

[1] Ksh ≤ 100 [2]101–500 [3] Ksh501–1,000 [4] Ksh1001–1500 [5] Ksh1501–2000

c)   How often do you pay for water?

[1] Daily [2] Weekly [3] Monthly [4] Others: Specify

d)   How much do you pay for water? ___Ksh

e)   How will you describe the current amount you spend on water?

[1] Low [2] Normal [3] High [4] Other: Specify

Water availability: (skip g and h if response for f is NO)

a)   From which infrastructure do you collect water for this household?

[1] Water kiosk [2] Yard tap [3] Personal connection [4] Other: Specify

b)   How much water do you collect in a day from this water source? ___Litres

c)   Is this water enough for your activities?

[1] Yes [2] No

d)   What household activities do you use water from this source for?

[1] Drinking [2] Cooking [3] Laundering [4] Bathing [5] Toilet flushing [6] Other: Specify

e)   How much water do you need for daily activities? __Litres

f)   Do you depend on other water sources or providers to meet your daily water use?

[1] Yes [2] No

g)   Which are the other sources you depend on?

[1] Groundwater [2] Surface water [3] Rain water [4] Cart pushers [5] Other: Specify

h)   What activities do you use water from these other sources and providers for?

[1] Drinking [2] Cooking [3] Laundering [4] Bathing [5] Flushing toilet [5] other: Specify

Time spent collecting water

a)   How far do you have to walk to collect water?

b)   How long do you have to wait at the water source to collect water?

c)   How many trips do you makes to the water point in a day?

Continuity of supply

a)   How many days in a week do you receive water?

b)   How many hours in a day do you receive water?

c)   Do you experience supply interruptions at your water point? [1] Yes [2] No

d)   How often do you experience these interruptions at your water point?

[1] Not often [2] Often [3] Very often [4] others: Specify

e)   What is the response to maintenance whenever it is needed?

 [1] Prompt [2] Delayed [3] If delayed Specify

f)   Which of the aspects of your water supply do you suggest need improvement?

[1] Pressure [2] Reliability [3] Maintenance [4] Cost [5] None [6] Other: Specify

g)   How would you rate the existing water supply service?

[1] Very Good [2] Good [3] Average [4] Poor [5] Other: Specify

Appendix D: Questionnaire for water point operators

Time spent collecting water

a)   How long do residents have to wait at the water source to collect water?

Cost of Water

a)   At what price do you sell water?

b)   What is the approved price?

Water Sales

a)   How many people do you serve in a day?

b)   How much water do you sell in a day/month?

Continuity of supply

a)   How many days in a week do you receive water?

b)   How many hours in a day do you receive water?

c)   Do you experience supply interruptions at your water point?

[1] Yes [2] No

d)   How often do you experience these interruptions at your water point?

[1] Not often [2] Often [3] Very often [4] others: Specify

e)   What is the response to maintenance whenever it is needed?

 [1] Prompt [2] Delayed [3] if delayed: Specify

f)   Which of the aspects of your water supply do you suggest need improvement?

[1] Pressure [2] Reliability [3] Maintenance [3] Other: Specify

g)   How would you rate the existing water supply service?

[1] Very Good [2] Good [3] Average [4] Poor [5] Other: Specify

LIST OF ACRONYMS

ADB	African Development Bank
DANIDA	Danish International Development Agency
DMM	Delegated Management Model
EU	European Union
GIZ	Deutsche Gesellschaft für Internationale Zusammenarbeit
GoF	Government of France
GoK	Government of Kenya
IFAD	International Fund for Agricultural Development
KEWASCO	Kericho Water and Sanitation Company
KEWASNET	Kenya Water and Sanitation Civil Society Network
KfW	Kredit fuer Wideraufbau
KIWASCO	Kisumu Water and Sanitation Company
KUAP	Kisumu Urban Apostolate Programs
LIA(s)	Low-income area(s)
MWI	Ministry of Water and Irrigation
NAWASSCO	Nakuru Water and Sanitation Company
NGOs	Non-governmental Organisation
NWMP	National Water Master Plan
NWSS	National Water Services Strategy
PEWAK	Performance Enhancement of Water Utilities in Kenya
PPDs	Pre-paid Water Dispensers
PPIP-WSS	Pro-poor Implementation Plan for Water supply and Sanitation
PPMs	Pre-paid Meters
RFID	Radio Frequency Identification
UN	United Nations
UN-HABITAT	United Nations Human Settlements Programme
UNICEF	United Nations Children's Fund

VEI	Vitens Evides International
WAGs	Water Action Groups
WASPA	Water Service Providers Association
WASREB	Water Services Regulatory Board
WSP	Water and Sanitation Programme
WSP(s)	Water Service Provider(s)
WSTF	Water Sector Trust Fund
WSUP	Water and Sanitation for the Urban Poor
WUAs	Water User Associations

LIST OF TABLES

LIST OF FIGURES

ABOUT THE AUTHOR

Akosua Sarpong Boakye-Ansah holds a Bachelor's degree in Laboratory Technology from the University of Cape-Coast, Ghana and Master's degree (cum laude) in Water Management from the IHE Delft Institute for Water Education (formerly UNESCO-IHE Delft Institute for Water Education). Between the periods of 2007 and 2015, Akosua worked with the Ghana Water Company Limited (GWCL) first as Chemist/Bacteriologist in a Central Water Laboratory and as a Process Chemist for the Barekese treatment works, in the Ashanti region of Ghana. In this capacity, she was responsible for the treatment and ensuring overall quality of water treated on the plant

In 2015, she started her PhD research with the Water Governance department of the IHE Delft Institute for Water Education. Her directly contributed to the project, Performance Enhancement of Water Utilities in Kenya through benchmarking, collective learning and innovative financing (PEWAK). This research focused on the conflicting demands of water utilities in their daily services to the urban poor – delving into the challenges water utilities face in ensuring equitable service provision in fragmented cities and understanding how the utilities try to address these challenges.

As a practitioner, she is experienced in drinking and wastewater treatment, water quality assessment and water distribution. As a researcher, she is interested in social, political, institutional and governance aspects of water services and been involved in researches in different countries including Ghana, Malawi and Kenya.

Journals publications

Publications directly linked to this thesis

Boakye-Ansah A.S., Schwartz K. and Zwarteveen M. The Appropriateness of Appropriate Technologies in improving access to utility water: Evidence from Three Kenya Cities (under review by the International Journal of Water Resources Development).

Akosua Sarpong Boakye-Ansah planned collected and analysed data from field research for this article. Klaas Schwartz and Margreet Zwarteveen supervised the work and contributed to the data analyses. Akosua Sarpong Boakye-Ansah drafted the manuscript. Klaas Schwartz and Margreet Zwarteveen critically revised the article for onward submission to the journal and during subsequent revisions.

Boakye-Ansah A.S., Schwartz K. and Zwarteveen M. (2020). Aligning stakeholder interests: How 'appropriate' technologies have become the accepted water infrastructure solutions for low-income areas. Utilities Policy, 66: 101081.

Akosua Sarpong Boakye-Ansah planned collected and analysed data from field research for this article. Klaas Schwartz and Margreet Zwarteveen supervised the work and

contributed to the data analyses. Akosua Sarpong Boakye-Ansah drafted the manuscript. Klaas Schwartz and Margreet Zwarteveen critically revised the article for onward submission to the journal and during subsequent revisions.

Boakye-Ansah A.S., Schwartz K. and Zwarteveen M. (2019). From rowdy cartels to organized ones? The transfer of power in urban water supply in Kenya. The European Journal of Development Research, 31(5): 1246-1262.

Akosua Sarpong Boakye-Ansah planned collected and analysed data from field research for this article. Klaas Schwartz and Margreet Zwarteveen supervised the work and contributed to the data analyses. Akosua Sarpong Boakye-Ansah drafted the manuscript. Klaas Schwartz and Margreet Zwarteveen critically revised the article for onward submission to the journal and during subsequent revisions.

Boakye-Ansah, A. S., Schwartz, K. and Zwarteveen, M. (2019). Unravelling pro-poor water services: what does it mean and why is it so popular? Journal of Water, Sanitation and Hygiene for Development, 9(2): 187-197.

Akosua Sarpong Boakye-Ansah planned collected and analysed data from field research for this article. Klaas Schwartz and Margreet Zwarteveen supervised the work and contributed to the data analyses. Akosua Sarpong Boakye-Ansah drafted the manuscript. Klaas Schwartz and Margreet Zwarteveen critically revised the article for onward submission to the journal and during subsequent revisions.

Other peer reviewed publications

Zwarteveen, M., J.S. Kemerink-Seyoum, M. Kooy, J. Evers, T.A. Guerrero, B.Batubara, A. Biza, **A.S. Boakye-Ansah**, S. Faber, A.C. Flamini, G. Cuadrado-Quesada, E. Fantini, J. Gupta, S. Hasan, R. ter Horst, H. Jamali, F. Jaspers, P. Obani, K. Schwartz, Z. Shubber, H. Smit, P. Torio, M. Tutusaus and Anna Wesselink (2017). Engaging with the politics of water governance, Opinion piece, WIREs Water, 2017, e01245.

Rusca, M., **Boakye-Ansah, A.S.**, Loftus, A., Ferrero, G. and van der Zaag, P. (2017). An interdisciplinary political ecology of drinking water quality. Exploring socio-ecological inequalities in Lilongwe's water supply network. Geoforum, 84, 138-146.

Boakye-Ansah, A. S., Ferrero, G., Rusca, M. and van der Zaag, P. (2016). Inequalities in microbial contamination of drinking water supplies in urban areas: the case of Lilongwe, Malawi. Journal of Water and Health.

Conference proceedings

Boakye-Ansah A.S., Schwartz K. and Zwarteveen M. Aligning stakeholder interests: how 'appropriate' technologies have become the accepted water infrastructural solutions for low-income areas. Poster presented at the 20th WaterNet/WARFSA/GWPSA Symposium, 31 October, 2019. Johannesburg, South Africa.

Boakye-Ansah A.S., Schwartz K. and Zwarteveen M. Risky Business? Supply of utility water to urban low-income areas: evidence from Kenya. Poster presented at the 9th International Water Association (IWA) Young Water Professionals conference, 25 June, 2019. Toronto, Canada.

Boakye-Ansah A.S., Schwartz K. and Zwarteveen M. Supply of utility water to urban low-income areas: evidence from Kenya Oral Presentation at the 3rd WASPA International Conference and Expo, 10 May, 2019. Nairobi Kenya.

Boakye-Ansah A.S., Schwartz K., Zwarteveen M. The (im)possibilities of improving access to utility water in urban low-income areas through service differentiation: Evidence from Kenya. Poster presented at the Sustainability and development conference, 10 November, 2018. Michigan, USA.

Boakye-Ansah A.S., Schwartz K. and Zwarteveen M. Assessing the impacts of pro-poor interventions on the performance of urban water utilities. The case of Kisumu, Nakuru and Kericho. Oral presentation presented at the 19th Waternet Symposium, 2 November, 2018. Livingstone, Zambia.

Boakye-Ansah A.S. Assessing the impact of pro-poor interventions on access to utility water in low-income areas: the case of Nakuru, Kisumu and Kericho. Oral presentation at the 2nd CDS conference: Critical Perspectives on Governance by Sustainable Development Goals: Water, Food and Climate, 25 June, 2018. Amsterdam, The Netherlands.

Boakye-Ansah A.S., Schwartz K. and Zwarteveen M. From rowdy cartels to organized ones? The transfer of power in urban water supply in Kenya. Oral presentation at the 18th WaterNet/WARFSA/GWP-SA Symposium, 27 October, 2017. Swakopmund, Namibia.

*Netherlands Research School for the
Socio-Economic and Natural Sciences of the Environment*

D I P L O M A

for specialised PhD training

The Netherlands research school for the
Socio-Economic and Natural Sciences of the Environment
(SENSE) declares that

Akosua Sarpong
Boakye-Ansah

born on 17 July 1983 in Asokore-Mampong, Ghana

has successfully fulfilled all requirements of the
educational PhD programme of SENSE.

Amsterdam, 5 November 2020

Chair of the SENSE board   The SENSE Director

Prof. dr. Martin Wassen   Prof. Philipp Pattberg

The SENSE Research School has been accredited by the Royal Netherlands Academy of Arts and Sciences (KNAW)

K O N I N K L I J K E   N E D E R L A N D S E
A K A D E M I E   V A N   W E T E N S C H A P P E N

The SENSE Research School declares that Akosua Sarpong Boakye-Ansah has successfully
fulfilled all requirements of the educational PhD programme of SENSE with a
work load of 52.2 EC, including the following activities:

SENSE PhD Courses

o Environmental research in context (2017)
o Research in context activity: 'Coordinating bi-weekly seminars for Integrated Water
   Systems & Governance (IWSG) Chair group' (2015-2017)

Other PhD and Advanced MSc Courses

o CERES basic training course: modules Research question and theory, Methodology in
   relation to research proposal and knowledge production, and presentation tutorials,
   Utrecht University (2016)
o Comparative epistemologies in development research (2016)
o A Practical course on the Methodology of fieldwork (2016)
o Qualitative data analysis for development research (2016)
o Comparative epistemologies in development research (2016)

Management and Didactic Skills Training

o Member of PhD association board IHE (2017-2018)
o Chair PhD Association Board (2018-2019)
o Supervising two MSc students with thesis entitled 'Serving the poor through pro-poor
   units: a case study in two Kenyan utilities' and 'Donor influence of water utilities
   implementation of pro-poor strategies; a case study of a water utility in Nakuru, Kenya'
   (2017-2018)
o Teaching in the MSc course 'Urban Water Governance' (2016)

Oral Presentations

o *From rowdy cartels to organized ones? The transfer of power in urban water supply in
   Kenya*, 18th WaterNet/WARFSA/GWP-SA Symposium, 27 October 2017, Swakopmund,
   Namibia
o *Assessing the impact of pro-poor interventions on access to utility water in low-income
   areas: the case of Nakuru, Kisumu and Kericho*. 2nd CDS conference: Critical
   Perspectives on Governance by Sustainable Development Goals: Water, Food and
   Climate, 25 June 2018, Amsterdam, The Netherlands
o *Assessing the impact of pro-poor interventions on access to utility water in low-income
   areas: the case of Nakuru, Kisumu and Kericho. 19th WaterNet/ WARFSA/GWP-SA
   Symposium*, 1 November 2018, Livingstone, Zambia
o *Supply of utility water to urban low income areas: evidence from Kenya*. 3rd WASPA
   International Conference, 10 May 2019, Nairobi, Kenya

SENSE coordinator PhD education

Dr. ir. Peter Vermeulen

CPSIA information can be obtained
at www.ICGtesting.com
Printed in the USA
LVHW101631240521
688348LV00013B/511